VGM Opportunities Series

OPPORTUNITIES IN
ENGINEERING
CAREERS

Nicholas Basta

Foreword by
R. A. Ellis
Director, Research, and Editor, *Engineers*
American Association of Engineering Societies

VGM Career Horizons
a division of *NTC Publishing Group*
Lincolnwood, Illinois USA

Cover Photo Credits
Clockwise from upper left: IBM Company; photo by Brownie Harris,
courtesy of General Electric; The Document Company, Xerox; National
Association of Professional Engineers/National Action Council for Minorities in Engineering.

Library of Congress Cataloging-in-Publication Data
Basta, Nicholas
    Opportunities in engineering careers / Nicholas Basta.
      p,   cm. — (VGM opportunities series)
    ISBN 0-8442-4591-7 (hbk.) — ISBN 0-8442-4592-5 (pbk.)
    1. Engineering—Vocational guidance. I. Title. II. Series.
TA157.B342   1996
620'.0023—dc20   95-31373
                                                      CIP

Published by VGM Career Horizons, a division of NTC Publishing Group
4255 West Touhy Avenue
Lincolnwood (Chicago), Illinois 60646-1975, U.S.A.
© 1996 by NTC Publishing Group. All rights reserved.
No part of this book may be reproduced, stored in a retrieval
system, or transmitted in any form or by any means,
electronic, mechanical, photocopying, recording or otherwise,
without the prior permission of NTC Publishing Group.
Manufactured in the United States of America.

5 6 7 8 9 0 VP 9 8 7 6 5 4 3 2 1

CONTENTS

The everyday impact of engineering. Profile of the profession
today. What is engineering? What engineers do. Engineers and
scientists. Engineering and technology. Engineers and the
world. Career options. Women, minorities, and engineering.

Career opportunities and goals. Areas of work. Education.
Employment in the private and public sectors. Engineers and
computers. Salaries and the intangible rewards.

Electrical/electronics engineering. Mechanical engineering.
Civil engineering. Chemical engineering. Mining,
metallurgical, and petroleum engineering.

Industrial/manufacturing engineering. Aerospace/
aeronautical engineering. Materials engineering.
Environmental engineering. Computer engineering.

ABOUT THE AUTHOR

For the past fifteen years, Nicholas Basta has worked as a business and technology journalist in New York, focusing on manufacturing, government policies, environmental activities, computer technology, and professional careers. He has been a regular contributor to *Chemical Engineering, Graduating Engineer,* and other publications. His books include *The Environmental Career Guide: Job Opportunities with the Earth in Mind* (John Wiley 1991) and *Major Options: The Student's Guide to Linking College Majors and Career Opportunities* (HarperCollins 1991).

Basta graduated in 1977 from Princeton University with a B.S. degree in chemical engineering.

FOREWORD

Engineering is by far the largest of the scientific and technical professions. Indeed, of all professions, only elementary and secondary school teachers outnumber engineers. Given the sheer size of the field and the enormous variety of attractive positions open to its practitioners, one might expect that it would not be too difficult to find good information about engineering careers, but in fact it is not easy to find accurate guidance to the field. The abbreviated material in encyclopedias and such references as the *Occupational Outlook Handbook* are not bad, but people pondering lifetime commitments to a career deserve much more.

Thus it is a pleasure to introduce *Opportunities in Engineering Careers* to its readers. This guide is *thorough:* It covers the full range of career-related concerns, from information on engineering positions and the people in them to practical tips about engineering colleges, licensing and professional registration, and other steps along the way. It is *even-handed:* It discusses drawbacks as well as advantages of the field and avoids hype about the most trendy specialties, a common failing of other materials about the engineering profession. Most of all, it is *wide-ranging:* It touches on a full selection of all the different roles and achievements of engineers at work.

The diversity of the profession is astonishing. Nicholas Basta rightly begins with traditional engineering: the electrical engineers who create computers, communication networks, and power systems; the mechanical engineers who produce every sort of machinery; and the civil engineers who design bridges, highways, and other kinds of structures and systems. But he also pays attention to the countless other applications of modern engineering training. More than ever engineers are moving into every area of the economy, including health care, finance and banking, and wholesale and retail trade. Moreover engineering training is increasingly being seen as the "new form of liberal arts," applicable to many kinds of jobs and particularly crucial for management.

The very nature of engineering—that it deals with leading-edge technology—means that the profession is constantly changing. The engineer in the twenty-first century will deal with new materials, constantly accelerating applications of automation, miniaturization of an unprecedented nature, and all sorts of other marvels. They also will deal with a multinational world in which work groups can draw on personnel located across the globe. For those able to deal with the science and math on which all technology depends, few career choices will be able to surpass engineering. Here is a handbook for that future.

R. A. Ellis
Director, Research, and Editor, *ENGINEERS*
American Association of Engineering Societies

INTRODUCTION

Do you like science? Would you like to translate that familiarity with science into faster computers, energy-efficient automobiles, spaceships, economical homes, better health care, cleaner water and air, and food for a growing world population? The path is before you—study engineering in college.

Engineering is the key that opens the doors of nearly all types of technology. In fact, most of the elements of our increasingly technological society are the product of engineers' work. The traditional image of the field is of an engineer examining a gigantic iron machine in a dirty, noisy factory. Today's engineers, however, move all over the country, from corporate headquarters to laboratories, spaceship runways, hospitals, even the halls of government. As technology has broadened its impact on how we live, it has created a wealth of careers for those who understand technology.

Technology provides the tools by which humanity copes with the world around it. Yet, it is a bad word to many people who believe that technology is devastating our environment. It is true that the world is different today than it was fifty or a hundred years ago. The main difference, however, is that there are nearly five billion people on the planet; the world's population has doubled at least twice during this century. Technology, today, means the difference between living and starving for millions of people.

Of course, technology also provides the tools to interact with our environment and be entertained by it. A typical home has televisions, radios, an automobile, a telephone, easily cleaned fabrics and materials—the list goes on and on. Engineering has a part to play in all these things.

Engineering Disciplines

This book will introduce the engineering profession to you and describe the many types of engineers at work today. Engineering is not a profession, or an academic program, that was "set" years ago. It is evolving every year. Chapter 2, The Common Elements of Engineering, describes some basic engineering functions and job possibilities open to all types of engineers. The classic engineering disciplines, most of which were established in the nineteenth century, are described in Chapter 3. These disciplines are:

civil
mechanical
electrical/electronic
chemical
mining, metallurgical, and materials

In the early years of the twentieth century—as the airplane, the automobile, and factory assembly lines changed the face of American industry—a new group of engineering disciplines arose. Later, after World War II, the advent of electronics, the computer, and nuclear power led to a diverse group of newer disciplines. These engineers are grouped in Chapter 4, Modern Engineering. They include:

industrial/manufacturing
aerospace
materials
environmental
computer

All these disciplines constitute the bulk of engineers being educated today. The smallest field, materials, welcomed 845 B.S. graduates in 1994, according to the Engineering Workforce Commission (Washington, DC). The largest field, electrical/electronic, added 15,935 graduates.

But there are many types of engineers. Chapter 5, Engineering Specialties, provides a summary of the more notable disciplines:

biomedical/biotechnology
agricultural
nuclear
marine/ocean and naval architecture
safety and fire protection
optical
automotive
textile
energy
heating, ventilating, and air-conditioning
systems/operations research

Chapter 6, Engineering Technology, covers the variety of programs and work responsibilities for engineering/industrial technology graduates, a field closely associated with engineering.

The last section, Chapter 7, Engineering Education, provides details on the best ways to prepare for engineering study, how to select a school, and how to get your career off to the right start.

Profiles

Interspersed throughout the book are a number of profiles of working engineers. These profiles are part fiction, part fact. They are created as a composite of real engineers, based on the author's fifteen years of reporting on the engineering professions. Names, of course, have been changed. The engineering field is diverse, and opportunities ever-changing, so it is not advisable to model your own career too closely after one of the profiles.

The actual work that engineers do is a vast mystery to most of us. It is rare for an engineer to be the subject of a television series or the hero of a novel. That's a shame because there is plenty of drama in designing and building a bridge or in testing a new aircraft. If nothing else, these profiles will give the flavor of engineering work.

ACKNOWLEDGMENTS

A considerable number of organizations and individuals provided assistance and advice in the creation of this book. They include The American Academy of Environmental Engineers; the Institute of Electrical and Electronics Engineers, Inc.; the American Society of Agricultural Engineers; the American Society of Mechanical Engineers; the American Society of Civil Engineers; the American Institute of Chemical Engineers; the American Society of Engineering Educators; the Institute of Industrial Engineers; the Society of Automotive Engineers; the Instrument Society of America; the American Society of Safety Engineers; the Society of Manufacturing Engineers; and the Junior Engineering Technical Society.

In addition to these individuals, I would like to add J. Robert Connor and Richard Ellis, of the Engineering Workforce Commission. Finally, I would like to remember the late Betty M. Vetter, founder and director of the Commission on Professionals in Science and Technology (Washington, DC). For decades her group gathered data and produced studies that helped form the vibrant engineering professions that exist today.

CHAPTER 1

ENGINEERING: GATEWAY TO THE WORLD

THE EVERYDAY IMPACT OF ENGINEERING

Take a look around yourself, no matter where you are sitting or standing as you read this. What do you see? You may see books, including this one that you are holding. Perhaps you see tables, chairs, and shelves if you are in a library. Maybe there is a couch, television set, carpet, windows, and walls if you are reading this book at home.

No matter what you see, you can be assured that an engineer was involved in designing or making it. Mechanical engineers helped make the machines that produced the paper this book is printed on; chemical engineers produced the ink of these words. Textile engineers manufactured the woven fabrics that make up your clothes; civil and materials engineers developed the paints, structural materials, and windows that make up the room around you. If you're in any kind of vehicle, many different engineers had a hand in designing and producing it.

These examples only scratch the surface of the activities and responsibilities of engineers. Engineering represents a group of skills that are central to modern life. Computers, spacecraft, telecommunications, and all the other forms of high technology are obvious fruits of engineering practice. Not so obvious are the water we drink, the air we breathe, the houses we live in, even the food we eat. Individuals with training in engineering can gain entry into practically every form of business and the arts that makes up our society.

Engineering offers the chance of lively, interesting work. More practically, it is one of the most reliable forms of employment. In recent years about half of the job offerings that campus recruitment offices report have gone to engineering students, even though they represent no more than 10 percent of the graduates. Starting salaries for engineers are invariably the highest of any given to graduates with bachelor's degrees. Also, it is rare for the unemployment rate of engineers as a group to rise above a couple of percentage points, even when the economy is suffering.

In recent years the quality and quantity of engineers have been matters of national concern. Although most workers (of all types, from presidents to plumbers) are not in manufacturing, that sector of the economy is the driving force that sets all the other sectors in motion. The majority of engineers work either directly in manufacturing or in a host of services that support manufacturing. In today's world, nations generally are no longer competing militarily (thankfully!); instead, their economies compete through trade. The U.S. manufacturing base is the foundation of our success in competing in this arena. When the economy suffers, one of the first checkpoints for national policymakers is the health of the engineering professions.

Besides its job security and its central position in the national economy, engineering often has opportunities for highly creative work. Most inventors have some type of engineering training. Think of the tremendous changes being wrought in our lifestyles by the introduction of the personal computer. Engineers were involved in the early days of computing and are at the forefront of developing everything from compact disks to the Internet for using computers today. Some engineers work in highly routinized jobs, checking the same product characteristics day after day. But others are more involved in finding new ways to build or make things and in solving pressing social issues.

Something that sounds as good as engineering must have a catch, right? Well, there is no denying that engineering study is hard work. Engineering students must contend with a lot of math and science. Regardless of the type of engineering that interests you, it is based on various scientific discoveries, and to be a good engineer, you must be familiar with science.

Most engineering students, and many working engineers, must fight their image as nerds. Some of the top college students, many of whom would make excellent engineers, choose to go on to law or medical school simply because lawyers and doctors have the reputation of higher status than engineers.

Another drawback is that although most American industries would cease to exist without engineers, those industries tend not to promote engineers as presidents or top executives. Backgrounds in finance or marketing are often preferred over engineering. This tendency is unusual. Both in West Germany and Japan, the two countries besides the United States that are renowned for their technological prowess, most major manufacturing companies are headed by engineers.

PROFILE OF THE PROFESSION TODAY

Most of these disadvantages, however, are ones of perception, not of reality. A great number of leading U.S. companies have engineering managers. Although engineering is not an automatic ticket to upper management, it is certainly no impediment. And the nerdy image of engineering is one that is rapidly fading away. With more and more engineering work involving teams of experts solving manufacturing problems, the time is passing when an engineer could be a lone operator in a laboratory, spending more time with a computer than with fellow workers.

Fortunately, the status of engineers in the United States is changing. In late 1989 the top executives at General Motors, IBM, and General Electric, to name a few, were engineers. And the global competitiveness that American manufacturers are developing is bringing the production-management role—usually run by the engineering staff—into the forefront. Better engineering means better manufacturing, higher quality, and more innovation—all the things that American industry needs to succeed.

Engineering used to be thought of as a white male club, but that time has gone, never to return. All the engineering branches are reaching out to women and minority groups. If you are in either category, engineering is opening its doors to you. There are professional organizations that offer support, and there are already many women and minority engineers who are rewriting the book on the American success story.

Finally, engineering is excellent preparation for tackling many of society's ills. The situation of the homeless cries out for a solution to the problem of affordable housing. The environment has become a national issue of the highest importance; engineering skills will be needed to make amends for the pollution of the past and to prevent new forms of pollution. Poverty is cured by jobs, and jobs are created when engineers develop new industries and tools. The problem of hunger can be confronted with more economical means of producing and delivering food.

This is a book to guide you in choices of engineering disciplines if you are already leaning toward engineering. And if you are not so inclined, it will show you why engineering might be your best choice. If mathematics and science are a snap for you to learn, engineering can be the best possible way to exercise that talent. Even if math and science are intimidating to you, don't automatically reject engineering. These hurdles can be overcome, and you will have a chance to apply a strong technical training in all sorts of interesting ways.

WHAT IS ENGINEERING?

Here is the formal definition of engineering, as espoused by the American Society for Engineering Education:

> Engineering is the profession in which a knowledge of the mathematical and natural sciences gained by study, experience and practice is applied with judgment to develop ways to utilize the materials and forces of nature, economically for benefit of mankind.

This is a definition that was surely argued over and amended many times by a large number of people. Therefore every word in it has a precise meaning. The key words, as this engineer sees it, are these:

"Mathematical and Natural Sciences"

No getting around it: engineering involves a lot of math and science. In practice, however, most engineers don't work with any mathematics above calculus, which many students learn in their last year of high school or first year of college. More math is taught—such as linear algebra, differential equations, and so forth—because educators want to be sure that engineers are well grounded in mathematics at the outset of their careers. Mathematics beyond calculus is also very important for engineers who go on to get a master's degree or doctorate. But don't be intimidated—many engineers are very successful in their work with no mathematics beyond algebra.

The sciences are another story. One way or another, all engineers are involved with sciences such as chemistry, physics, geology, or materials—but not all sciences, and not all the time. This is a key difference between the profession of engineer and that of scientist.

"Study, Experience, and Practice"

Engineering combines art and science, and "art" means that the engineer depends on many things that haven't been reduced to mathematical equations. Engineers often depend on rules of thumb or calculated guesses. They work with approximations, with unknowns, and with their intuition and judgment. Nevertheless, the work must be on target—very often, lives are at stake in a bridge or aircraft design. That's why these structures are built with a margin of safety and are thoroughly tested before use.

"Economics"

This may be the most important, or at least the most distinctive, aspect of engineering. A doctor will spend whatever it takes to heal a sick person; a lawyer can continue fighting a cause in court until funds are exhausted. But an engineer, every day of his or her work life, is constantly battling to produce goods more efficiently, to save energy, to conserve resources and the environment, and to reduce wear and damage. The difference between a failed product and a wild success can be as little as fifty cents in production costs. Engineers are continually confronting economics.

The importance of economics led to this informal definition of engineering: engineering is doing for one dollar what any darn fool can do for two. Although money and economics may seem a boring focus in one's work, they are truly the exciting aspect of engineering. The next time you buy a can of soup, realize that it costs less today than it did 150 years ago when canned food was invented. And if you use a personal computer, realize that it has the same computing power, in a box costing a couple of thousand dollars, that a room-sized monster machine costing millions of dollars had in the 1960s. Smart engineering made the difference.

WHAT ENGINEERS DO

What do engineers do? They build bridges, design aircraft, run power plants and factories, and get ore from the ground. All these things are well known, but they are not the only tasks performed by engineers. Engineering work is as varied as the individuals who practice it.

Engineers not only design aircraft, for example; they also build them, test them, and fly them. That includes everything from satellites to blimps, gliders, or rockets. The first man on the moon, Neil Armstrong, was an engineer.

Engineers build bridges; they also build tunnels, highways, dams, airports, and docks. For living or working space, they design and build homes, offices, and factories. And to those who say that engineers only destroy nature by building things, one can reply that engineers are also involved in preserving wetlands and shorelines, restoring forests, and cleaning up dumps.

Engineers not only run factories; they are also developing robots that eliminate boring, repetitive work. They help establish innovative work practices such as self-managed production lines or quality circles—two ways to increase the productivity of workers and the quality of the products. They help develop production techniques that reduce pollution and raise efficiency.

And finally, to those who say that engineers work only in smoky factories or on noisy construction work sites, one could reply that engineers also help artists create new visual forms and musical sounds. They help design new entertainments at amusement parks or video arcades and new sports equipment for the use and enjoyment of the Olympic athlete or the weekend hacker. Engineers are in government, education, charities, and community work. Yes, many engineers work on the factory floor; but many also work in laboratories, offices, on ocean platforms, in the mountains, at the shore, or on city streets.

If there is one dominant message in this book, it is that an engineering education doesn't lead to only one or two types of careers or work environments. Engineering is a gateway to a huge, diverse array of opportunities.

ENGINEERS AND SCIENTISTS

Many young people who excel in the sciences aren't sure whether a career as an engineer or as a scientist is the best future for them. In many ways, the two careers are similar. Many engineers, especially those who get an advanced degree, do exactly the same type of work as scientists with advanced degrees. They can both work in laboratories, running experiments and analyzing data to develop fundamental rules or principles about how nature works. Conversely, many scientists start their careers in a corporate laboratory or as a quality manager in a production line and then develop into factory managers or administrators.

Some sharp students, having noticed the commonality between the two fields, have decided to study in one field as an undergraduate and in the other as a graduate student. The switch from engineering undergraduate to scientific graduate is somewhat easier than the reverse because engineers take more math than most science majors do. However, it can be done either way.

There are, of course, differences between the two. Most college-level science programs are designed to prepare the student for work in a laboratory or for graduate school. Engineering programs are designed to prepare students for work in business and industry, with opportunities ranging from design to production to sales. This diversity is one of the reasons why there are so many more employment opportunities for engineering students immediately after graduation.

Methods

On another level, the difference between engineering and science is one of philosophy. Scientific work is a search for the truth; engineering work is a search for what is practical. The scientific method, formally defined, is a way to determine the truth or accuracy of a principle by performing experiments that either confirm or deny that principle. Thus, scientific work tends toward conducting lengthy sets of experiments, studying reference materials in libraries, and writing presentations for publication in science journals.

The engineering method has been defined by Billy Vaughn Koen, a professor of mechanical engineering at the University of Texas (Austin), as:

> The strategy for causing the best change in a poorly understood or uncertain situation within the available resources.

This phrase demonstrates both the power and the limitations of engineering work. Engineers are constantly dealing with uncertainties and often don't have the time or money to obtain an understanding of the scientific principles involved. Thus, engineering work tends toward finding what works for a given situation in the shortest possible time and then using that solution until another situation arises. Engineering work is a striving for constant improvement. Rather than going to a library to gain insight on a topic, an engineer is more likely to check with workers on the production line, or with consumers of a product.

Working Conditions

It isn't easy to take abstract principles and understand what work one will be doing in the future. Here's a concrete example from the electronics industry. In the late 1940s, three scientists at Bell Laboratories, while trying to establish some new properties of silicon, developed the transistor, a mixture of silicon and germanium. It had such dramatically new properties that a Nobel Prize was awarded the researchers, and a new "solid-state" age of electronics began. (Previously, electronic devices were based on vacuum tubes; a complex mixture of glass bulbs, wires, and electric power.)

By comparison, in the mid-1980s a team of engineers at Intel Corporation—the Santa Clara, California, microelectronics producer—were charged with developing the 80486 microprocessor, a circuit chip that would power the next generation of computers. The team had to devise a way to cram 250,000 transistors onto a chip the size of a fingernail, and they had eighteen months to complete it. Almost simultaneously, another team at Intel was being set up to design what came to be known as the Pentium chip, with over two million transistors, as the replacement for the 486. Each team was aware of the other, but each had different goals and timetables. The 486 team could use one type of materials and fabrication techniques, while the Pentium could use others that were not as well established commercially.

Bell scientists had essentially no timetable to meet because no one was sure whether the things they were trying to accomplish could actually be done. When the scientists made their discovery, they knew that they had developed a fundamentally new way of working with semiconductor materials. They had the satisfaction of knowing that their discovery would eventually change the way electronic equipment was made.

The Intel engineers, on the other hand, had a strict timetable to meet. Their invention (which has generated several patent applications) met predetermined performance goals. They had the satisfaction of knowing that their work would result in sales of new computers and microelectronics for their employer; it would also

help the company's customers perform their work more efficiently. But they knew that their product would not live forever
because one day the 80486 chip would be replaced by the Pentium
chip.

Most scientists, and nearly all engineers, work in industry. But
proportionately more scientists work in academia, teaching, and
research. If you are happy in a school environment, there are more
opportunities with a science background than with an engineering
one. Conversely, an engineering background is more likely to result in a job in industry. It's up to you to decide.

ENGINEERING AND TECHNOLOGY

Have you ever helped out in repairing a car, or opened up a radio and tried to take it apart? In figuring out how machines work,
you can also figure out how to fix them. Keeping machines running is the fundamental task of technicians, and you may find that
your interests lie more in this line of work than in engineering.

What's the difference between these two? Engineers, after all,
also help keep machines running. The key difference isn't the type
of machine, but in the approach taken to machinery by engineers
and technicians.

Technicians, fundamentally, use and repair the machines that
engineers develop. A technician will take an existing machine and
apply it to some task. Or, the technician may need to discover why
a good machine suddenly stops functioning.

A Chemical Factory

In a chemical factory, for example, samples of the chemicals
that are being produced are taken out of the production vessels periodically. These samples are sometimes brought to a central labo-

ratory where a technician takes the sample and runs it through an instrument called a chromatograph. This device has the ability to separate chemicals in the sample mixture and measure how much of each chemical the mixture contains. The technician starts up the chromatograph, checks that it is operating properly, puts the sample through, and then analyzes the results. He or she then writes a report describing the components of the mixture and delivers the report to the people running the production process.

An engineer, on the other hand, designed and manufactured the chromatograph. Engineers were also at the chemical factory earlier, when it was built, to help design the process and see that it was constructed properly. Even earlier, engineers were involved in developing the very chemicals that are being produced and in figuring out how to produce them economically. At the present time, one or several engineers are involved in making sure that the process is running as efficiently as it can. It may be an engineer, for instance, who receives the report from the technician and decides that the process must be changed.

This example illustrates how an engineer may set up the tools and equipment for a manufacturing process but leave that process in the hands of technicians to run and maintain. A technician's work can require a high level of sophistication; in some cases, a technician knows more about how to use a machine than the engineer who designed it. Then, too, sometimes the machine is so complex that an engineer is needed to run it. Some machines, especially ones that are newly invented, require complicated tests and analyses even before they perform the intended task.

Maintenance Work

Maintenance work is another way of illustrating the differences between engineers and technicians. Often engineers who are very knowledgeable about a certain type of equipment will know very little about how to maintain that equipment. Computer maintenance, for example, can require a sophisticated understanding of how electronics work and how to solder components together. But many computer engineers work only with blueprints or circuit diagrams and have never held a soldering gun in their hands. However, when a maintenance technician has performed all the tests or repairs he or she knows and a system still doesn't work, an engineering team can be called. A problem that can't easily be repaired may indicate a fundamental flaw in the equipment's design, and the engineers will be best equipped to find it.

Training and Job Opportunities

In terms of training, many technicians are required to take only several months' worth of vocational courses after high school. Experience is a valuable commodity; a technician who has worked for several years is paid more than an inexperienced one. Many companies, especially those involved in computers, telecommunications, or heavy equipment (airplanes or earth movers) provide these training courses themselves. The engineer, of course, must have a college degree.

Depending on the individual, an engineer can often move rapidly into management positions and continue climbing up the corporate ladder. Technicians, on the other hand, have more limited possibilities for promotion.

If you think you have a strong mechanical aptitude, would you prefer operating and maintaining equipment or designing and building equipment? Many people aren't sure. One way to find out is to obtain technical work before or during college. The military offers many opportunities for technician's work, and some of this experience can be transferred directly to a job in the private sector.

Job demand for both types of workers is high today. The engineer tends to have a set of skills that can be transferred from one type of technology to another, while the technician's training is usually specific to one type of machinery or instrumentation.

ENGINEERS AND THE WORLD

Engineers are the inventors and implementers of technology. It is common to hear that American society—indeed, the world—is becoming more technological. What does this mean?

In a major study published in 1988, the Congressional Office of Technology Assessment (OTA) considered this question. OTA is a federal agency controlled by the U.S. Congress that examines political issues in a technological context. It regularly issues reports on new technologies in areas such as health care, communications, pollution control, or education. In the 1988 report, entitled *Technology and the American Economic Transition,* OTA looked at how Americans live and work. They found a complex set of networks, an interconnected web of businesses, economic activities, and people. One example cited goes as follows:

> What could be more basic than frozen pizza? A man cooking a frozen pizza in a microwave oven cares about what the pizza costs, how it tastes, how its preparation fits into his increasingly harried lifestyle...Consider a likely chain of events that culminated in the pizza. Knowledge about the health effects of food came from a TV talk show, and information about a sale on pizza came from a newspaper ad. Wheat for the pizza crust was grown in Kansas using sophisticated seeds and pesticides. The pizza was assembled automatically and wrapped in materials that are themselves the product of considerable research. The pizza was probably purchased at a grocery store where a clerk passed it over a laser scanner, which entered data into a computer and communication system designed to adjust inventories, restock shelves, and reorder products. This system in turn made it possible to operate an efficiently dispatched transportation system, placing a premium on timely and safe delivery...

The point OTA is making in this example is that even when considering something like food purchased in a supermarket, the marks of technology are all around. Each network on which we depend for our well-being has technological elements. Engineers are actively involved in all of these elements.

The networks OTA analyzed are:

food
housing
health
transportation
clothing and personal care
education
personal business and communication
recreation and leisure
defense
government activities (besides defense)

If any of these networks surprise you by their presence, read on; you will see how engineering is involved. If you have already targeted one of them as the objective of your career, this book will show you how to get started.

Concern for the Environment

Even though most of these networks represent some type of business, it would be a mistake to believe that all engineers work in business. A good example is concern over the environment. Today's newspaper headlines continually trumpet new environmental worries: acid rain, hazardous waste, workplace safety, global warming, polluted water, and municipal garbage. There is dramatic debate today over humanity's place in the world and the condition of our planet, which we will pass on to the next generation. Some people believe that the answer is to turn the clock backwards—to reduce the technological complexity of modern life and to live more simply and in better harmony with nature. Others argue that the only answer to the environmental problems caused by technology is more technology. This debate is a philosophical and cultural one and will probably continue for many years to come. In the meantime, however, something has to be done, and engineers are the ones doing it.

WATER TREATMENT

Examples of this work are all around us. Most cities have a water treatment plant to remove wastes before water is discharged to nearby rivers, lakes, or seas. Developing these treatment plants was one of the first activities of civil engineers around the turn of the century. Today, hazardous wastes—the by-products of industrial production—are a critical concern. Engineers have developed methods, based on water treatment and oil-drilling technology, to inject nutrients deep underground. The nutrients are used by organisms that are capable of destroying harmful pollutants.

GARBAGE DISPOSAL

Another example can be seen in garbage disposal. Years ago the disposal practice of most cities was simply to dump it in the closest, most convenient location. Today, however, such landfills are being engineered to have relatively impermeable walls so that the dangerous materials in garbage do not escape into the environment. However, the amount of readily available landfill space is declining, so environmentalists are pursuing efforts to recycle much garbage back into commerically useful materials. Once the suitable materials are found in garbage, they have to be extracted, purified, and reprocessed. Engineers are involved at every step of the process.

Many people have looked at the vast mountains of garbage our society generates and asked why a more effective use and disposal of it wasn't possible. The result has been an increasing reliance on recycling—restoring throwaway items to something of value.

International Opportunities

In a profound sense, engineers are involved in helping sustain people around the world. In underdeveloped nations, the essential elements of life—water, shelter, transportation—are provided through the efforts of engineers. Advanced technology provides some interesting solutions to tricky problems. For instance, in sub-Saharan Africa, how does one keep pharmaceutical compounds that must be refrigerated? You can't simply put them in a refrigerator because an electric power station may be a thousand miles away. And you may not have the fuel available to run a portable generator. The solution is to use a complex synthetic mineral called zeolite that can use solar energy to provide cooling.

Another example can be seen today in the People's Republic of China where telephones are hard to come by. In Beijing, the capital city, it is common to see officials with cellular telephones, high-tech versions of this essential tool that only in the past decade has become common in the United States. Because underdeveloped countries like China lack the utility infrastructure that has been built up in the United States and Europe, a wireless system such as cellular telephones is more practical. In this way, the business and government community can take advantage of the latest technology, even when the rest of the country is very backward.

Headlines are being captured today by young engineers developing new "multimedia" entertainment, which combines sound, video, and text on a PC screen.

CAREER OPTIONS

How about the fun side of engineering? It does exist, even if you aren't aware of it. For example, a booming business today is the construction of amusement parks, complete with stomach-wrenching rollercoasters. These complex machines, as well fas many other rides, require extensive structural design work by engineers. You will also find engineers at work in rock-and-roll studios, concert halls, movie sets, sports arenas, and beaches.

Even before all the various engineering specialties have been described, you can see that there are dozens of different engineering functions, job descriptions, and opportunities. It's hard to decide that you want to be an aerospace engineer, a materials engineer, a computer engineer, or whatever. The solution to this problem is simply to make a decision first to be an engineer. Later, you can decide on the specific type.

Selecting a Specialty

There are many reasons, practical and emotional, to delay selecting a specialty. At nearly all college engineering programs, you don't select a major until the end of your freshman year or during your sophomore year. Taking college-level courses as a freshman, talking with faculty and fellow students, and thinking freshly about who you are and where your interests lie—all these things help guide one's decision. There's no need to rush into it now.

A second reason for waiting has to do with the nature of engineering work. Because engineering is a broad profession, there are many different types of work, even within each engineering discipline. Several of the engineering disciplines—especially mechanical, industrial, and electrical engineering—provide entry into nearly every type of manufacturing business, government, research, or other types of organizations. If you are a mechanical engineer, for example, obviously you could work in a field like the automotive industry. But you could also find opportunities in electric utilities, in aerospace firms, in government laboratories, and in other industries.

A third point is that, even within one type of industry, there is great variation in job responsibilities. At factories you could work in design, production, quality control, maintenance, or plant management. At the corporate headquarters, you could work in sales or marketing, business management, administration, or research and development.

A fourth and final point is that over the course of a career, engineers can do many different things. There are engineers who spend their entire career happily in one department of one company, and others who move all around the corporate ladder. There are engineers who start their own companies, or who work for themselves, as consultants, and never occupy a company office.

Employment Outlook

There are major differences in career prospects among engineering branches, and the purpose of this book is to point them out to you. At any given time, the job demand varies, depending on the condition of the economy. A look at Figure 1 illustrates this point. These data, from the latest forecast computed by the U.S. Bureau of Labor Statistics, show the size of various engineering branches as of 1992 and the projected growth through the year 2005. BLS does not publish data on occupations with fewer than 25,000 practitioners, so don't worry if your favorite discipline is not listed here.

The most important point is to make a commitment to engineering now and choose the specific type of engineering career later. Even in looking at a projection for the year 2005, bear in mind that by that year you may be five or so years into your career; that leaves another thirty-five years at least to your working life! The key, of course, is to be an engineer—some kind of engineer. The rest is up to you.

WOMEN, MINORITIES, AND ENGINEERING

Special mention deserves to be given to the people who traditionally have been ignored or excluded from the ranks of engineering: women and minority groups. Today, engineering is wide open to minority groups. Many engineers of minority ethnic backgrounds are highly successful and have founded companies worth millions of dollars. Among persons of Asian-Pacific heritage, in fact, proportionately more are engineering students than their fraction in the United States population as a whole. The problem for other minority groups, however, is that there are not enough members in the educational system. The high requirements for a

Figure 1. Engineering Job Growth.

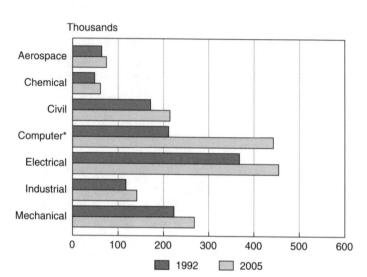

Source: Bureau of Labor Statistics.
*Computer scientists and engineers.

good high school education are an obstacle for some underprivileged inner-city children. The cost of a four-year college education adds to the difficulties.

However, there has been considerable progress in the entry of both women and minority groups into the ranks of working engineers. Figure 2 shows recent data for these groups, in terms of numbers of baccalaureate degrees awarded. Among minority groups, Asian-Pacific engineers are actually "over-represented," in the sense that their proportion is approximately twice what it is in the general U.S. population. Women's representation rose swiftly in the early 1980s, then dropped, but began picking up again in the early 1990s. Currently they represent about 17 percent of the engineering class.

The number of African-Americans, Hispanics, and American Indians in engineering started at exceedingly low levels and has grown slowly. There are many outreach programs that help sustain minority students in engineering. These include scholarships and intensive training sponsored by the National Action Council for Minorities in Engineering (NACME) and internships and financial support from leading corporations. About 50,000 non-Asian-Pacific minority engineers have won degrees since 1973. Many more will be needed in the future.

America needs more engineers, and in the face of a smaller number of college-age students of all races in the 1990s, America must get more engineering candidates from women and minority groups. Here is how the Commission on Professionals in Science and Technology, a Washington, DC, public-interest group, assessed the situation in 1989:

Figure 2. Minority and Women B.S. Degrees.

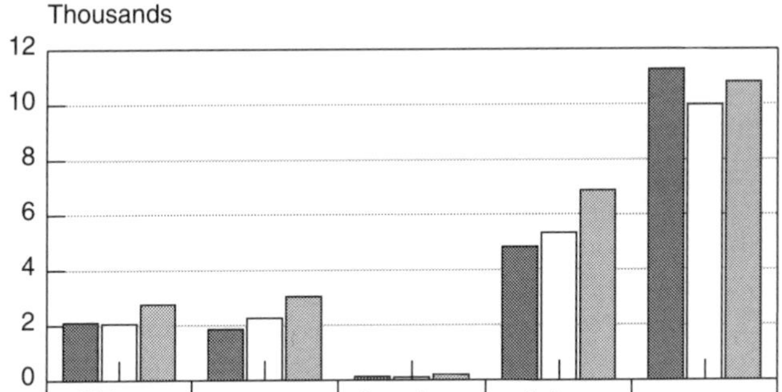

	African-American	Hispanic	American Indian	Asian-Pacific	Women
1986	2.114	1.864	0.129	4.824	11.264
1990	2.062	2.272	0.103	5.328	9.977
1994	2.769	3.045	0.207	6.881	10.800

■ 1986   ☐ 1990   ▨ 1994

Source: Engineering Workforce Commission.

...In science and engineering particularly, [women and minorities] are needed because the nation faces a potential shortfall of considerable magnitude in some of these areas, and white males cannot continue to fill all of the national needs by themselves.

The report, entitled *American Minorities in Science and Engineering,* examined the situation of engineers and scientists with Ph.D.s in particular detail. These college graduates will form the core of America's research efforts in the twenty-first century. Today, nearly half of all graduate students (master's and Ph.D. level) are foreign nationals—students from abroad who study here and often return to their homes after graduation. The report notes:

In 1988, when the size of the age-thirty population was at its peak,... American universities awarded doctorates in natural science and engineering to 7,455 American citizens, including 523 (7.2 percent) awarded to minorities. The National Science Foundation projection indicates positions for about 18,000 new Ph.D.s in natural science and engineering in 2004, when the size of the age-thirty population reaches bottom [around 3.2 million]. Where are they to come from?

The report concludes:

Half of our children are girls. A growing third are members of minority groups, and one sixth are...both female and minority. Put another way, only one third are white, non-Hispanic boys. The world's leading democracy cannot afford to depend on only one third of its population for leadership in science and medicine, in law and politics, and in all other fields, nor can it afford to have large segments of its youth ignorant of both the facts and processes of science.

Doors are opening wide for women and minority groups in engineering. Career prospects are outstanding. Think hard about yourself and your career aspirations before you pass over the opportunities that engineering represents.

CHAPTER 2

THE COMMON ELEMENTS OF ENGINEERING

This chapter will describe elements that nearly all engineering disciplines share. Most engineers go to school for four years, earning a bachelor of science in engineering (B.S.E.) degree or an engineering technology degree, and then get a job. But some go on to earn a master's degree (M.S.E.), and a few don't stop until they have a doctorate (Ph.D.).

Regardless of their specific degree, engineers can usually find jobs in design, research, technical services, and production. Management opportunities come with experience. Then there are a wide variety of fields from which to choose: manufacturing, construction, agriculture, business services, and government. The pay varies according to degree level, experience, the industry, and, of course, individual performance.

CAREER OPPORTUNITIES AND GOALS

Believe it or not, a great many people who study engineering during college end their careers in a field other than engineering. The reason for this is that over the course of a career, many engineers move into management, or into other lines of work altogether.

This isn't necessarily a bad idea. Each individual, during his or her working career, is usually making choices about whether to continue in an existing job, consider a promotion, or go to another employer. And in many lines of work, having an engineering degree demonstrates strong problem-solving skills that are useful in many other areas. Sometimes, unfortunately, a person finds future career growth stopped simply because there isn't a higher position to move into without leaving engineering work. The point here is that getting an engineering education is a way to open doors to a variety of career opportunities.

One of the first things you should do when considering engineering is to evaluate your long-term goals. You don't have to make a solemn commitment to the goal you choose now. But you should try to get a feeling for whether your long-term preference is research, production, design, or management. This decision will dictate how far you should plan on going in your education and will help guide your choice of engineering disciplines.

AREAS OF WORK

In the following chapters, there will be many details about the specific work in which various engineering specialties engage. Nearly all engineering work, at least early in one's career, falls into one of several categories.

Design

Design is probably the most common type of engineering work. Design simply means taking the knowledge of materials, processes, systems, and nature that one learns in college and adapting it to machines, equipment, structures, production methods, and government regulations. To look at an automobile, for example, is to see the efforts of thousands of design engineers who were instructed to improve a component or part on the basis of new information. It would be intimidating for one person to design each of the thousands of components of an automobile from scratch, although projects as complex as this have been done. More often, there is a large body of existing knowledge about why something was made a particular way. The design engineer becomes familiar with this body of knowledge through experience and then tries to adapt the component or structure to a new situation.

This process may seem dull or uninspiring, but in fact it is one of the most creative and exciting aspects of engineering work. When there is something truly new—a new type of plastic, or a new microchip, for example—the engineer must envision how those things will be used. Many of the best engineers consider themselves close to artists because they must envision things that have never existed before.

Production

This is probably the second largest area of work for engineers. Production is simply making whatever product a company is responsible for, as often as desired, and with the level of quality demanded by customers.

In a typical factory, there are many types of equipment doing something to materials or objects as they pass along a line. In a food-processing plant, for instance, grains of wheat can be successively ground to flour, mixed with various ingredients, shaped into cookies, baked, packaged, and shipped. In a steel mill, iron and other metals are melted in a furnace; poured into molds; shaped by rolls into sheets, plates, or bars; treated by heat or chemicals for purity; and put on trucks or railcars to be shipped to another factory. There an equally complex set of steps turns the steel into an appliance part or auto chassis.

Production engineers are responsible for keeping the assembly lines running. This involves an understanding of the product's function and specified quality. In addition the production engineer must know how the various machines that handle the product are operated, so they can be fixed if they break or begin to turn out faulty products. The capabilities of workers must also be taken into consideration. If a machine is being operated incorrectly or inefficiently, that will also affect the final product. Finally the production engineer must keep an eye on the cost of all the production steps. It may be worthwhile to spend a lot of money to improve some step in the production process if the improvement results in lower operating expenses or higher quality.

Production engineering is a fairly direct route to corporate management because no manufacturer can continue to exist if the products it makes are shoddy or cost too much. Production engineers can move from responsibility for one component in a manufacturing process to the entire plant and then to groups of plants.

Construction

Construction, or civil, engineers use many of the same skills as production engineers. One difference is that construction engineers work on buildings, highways, and other permanent structures that are often one-of-a-kind. Like production engineers, construction engineers must understand the final product, the necessary equipment and worker skills, and the economics of the project.

Research

Research done by engineers is often the same thing as research done by scientists. Something is discovered in nature or in the real world, and the engineer or scientist brings that discovery into the laboratory and tries to find out why it occurs.

There is pure research performed by engineers—usually teachers at colleges and universities—that may have no immediate application in the real world. This research is similar to a scientist discovering a new type of star, or a new mathematical equation, and feeling a sense of accomplishment for that discovery. But most research engineers, especially those working for private industry, engage in something called "applied" research. In this case, the engineer is trying to find a way to accomplish a specific objective that his or her employer will then turn into a marketable product or service. For example, a research engineer trying to find out why the turbine blades in jet engines wear out might discover that a different alloy works much better than existing alloys. That discovery is then passed on to other engineers (sometimes called design engineers, sometimes development engineers) who will figure out how to make the turbine blade out of the new material.

Not all research engineers have a doctorate, but many of them do, and having that level of training will open many laboratories' doors. There is also a fairly steady interchange between research and design. At some companies a research engineer will discover a new principle or capability in a laboratory and then follow that discovery in each step through development, design, production, marketing, and sales.

Technical Services and Consulting

Many engineers find lucrative employment by offering nothing more than their judgment, written up in a report, just as a lawyer advises a client to write a contract a certain way, or a doctor advises a patient to follow a certain diet. The product that the consulting engineers sell is their understanding of the technical details of some issue.

Many consulting engineers work in some aspect of the construction industry. A client will come to the engineer saying, "We need to reduce traffic jams on our roads." The engineer will observe the situation, think about the underlying problems, and then offer a set of recommendations. The cost of implementing these recommendations is an important element of the final report. In another application many consultants are being hired today as "systems integrators" for computer networks. A big company will create a list of preferred vendors for the computer system and then turn the list over to the systems integrator. That person or firm will then figure out the best arrangement of components for achieving the customer's goals and provide a report of recommendations. Sometimes, the consultant will oversee the installation.

Technical services and consulting are wide-ranging and very dependent on the entrepreneurial skills of the engineer. Many engineers don't consider a career in consulting or technical services until they have several years' experience in some relevant type of work. Many large corporations, however, employ technical services engineers to help their customers solve problems that might occur in the use of the company's products.

Sales, Marketing, and Product Management

Traditionally engineers have been stereotyped as tongue-tied, shy people who are happier dealing with machines than with other people. That image is disappearing fast.

Today, because so many products that we buy or use are technologically complex, only engineers or similarly trained people are capable of explaining their functions and advantages to a customer. Consider the difference between a horse-drawn wagon (the usual way to get across America a century ago) and an airplane (the usual way to do so today). Back then, customers went to the wheelwright and explained the length of their trip, the number of people involved, and how much money they had to spend. The wheelwright, who made three basic types of wagons, told the customer what was available, how much it would cost, and what options could be added. Today, an airplane buyer, such as an airline, needs to know fuel efficiencies, operating ranges, noise levels, maintenance standards, weather patterns, instrument capabilities, passenger loads—the list goes on and on. Having an engineering degree isn't the essential background for making sales such as these, but it certainly helps.

Sales as a career requires good communication skills, the ability to get along with people, persistence, and technical know-how. Technical salespeople need to be able to translate the needs and requirements of the customer into the specifications for the appropriate product. Because this experience is so vital to a manufacturer's success in the marketplace, sales careers often lead to marketing management positions. In many cases the experienced sales engineer will be put in charge of developing a new product and coordinating the campaign to get it out into the marketplace. These product managers are key players in most high-tech companies.

EDUCATION

The majority of engineers earn a B.S.E. degree and then go into industry. Undergraduate programs for engineering are accredited by the Accreditation Board for Engineering and Technology (ABET), which periodically reviews the curriculum and teaching facilities of colleges before renewing their accreditation. Many states require that engineers in certain functions (such as public works construction) be graduates of an accredited program.

At the behest of the professional engineering societies, ABET sets the minimum number of specific courses that engineers take. It also specifies such things as laboratory courses, computer programming, and design projects. These courses then become the required curriculum of college engineering departments.

Even though formal study stops for most engineers at the B.S. level, education continues. Most employers have some type of classroom education for new employees. And many engineers find it helpful to take continuing-education courses during their career to learn about new areas of technology.

About 20 to 30 percent of engineers, depending on the type of discipline they are in, go on to earn a master's degree. Many engineers consider the master's degree the necessary final step for a fully trained engineer. It is quite common for students to take a full-time job and attend night school. Silicon Valley, the area south of San Francisco that is the heart of the electronics industry, came into being, in part, because of the strong base of good engineering schools in the vicinity. These schools served as an anchor for high-tech companies and provided a place where hired engineers could continue their education.

But many engineers can now continue their education in remote locations by gaining access to videotaped courses and computer-networked classes. As you will see in the following pages, many engineering specialties exist primarily at the master's level, so if you want to be fully prepared for working in these areas, the master's program is advisable. Usually M.S.E. graduates can expect to earn annually about $2,000–$4,000 more than B.S.E. graduates when they start working. But this difference fades over time as individual performance becomes apparent.

A substantial number of engineering graduates combine their engineering degree with a master's degree in business administration (M.B.A.). The engineering/M.B.A. combination is a powerful one for engineers who expect to move into management or to start their own businesses. So-called "techno-M.B.A.s" have the skills to run businesses, especially those that involve highly technical goods and services.

The doctorate degree (Ph.D.) is obligatory for engineers who expect to teach at the university level. Many research groups—either in private industry, at federally funded national laboratories, or at research foundations—prefer Ph.D.-level engineers. Annual pay can be $4,000–$10,000 more than that of a B.S.E. graduate.

A final element in the educational environment is the question of professional licensing. This license is earned by passing two tests and gaining several years' experience in actual engineering work. Only about 15 percent of all engineers take the time to get this license, and most of these are in the civil engineering profession where government regulations require licensed engineers to review blueprints.

EMPLOYMENT IN THE PRIVATE AND PUBLIC SECTORS

According to data on graduating classes, well over one-third of the job offers engineering students receive come from manufacturers. This is no surprise because manufacturing is the sector of the American economy most closely tied to technology.

But this doesn't mean that the remaining sectors of the economy are closed off to engineers. In all likelihood most students head toward manufacturing initially because that's where they find the highest salaries. But there are well-paying jobs in other sectors, and, if salary isn't an issue of high importance, practically the entire employment arena is open to the engineering graduate.

Engineers are attractive job seekers because their academic training prepares them to handle numbers easily and to approach work with a positive, problem-solving attitude. The question sets and tests that engineering students take during school are not just a means of getting grades; in many ways, these test questions are comparable to the problems engineers encounter in the real world.

Government service is another significant source of jobs for engineers; the government is perennially short of engineers. The main employment areas in the federal government are the Departments of Defense, Energy, the National Science Foundation (which helps run the network of national laboratories), and agencies such as the Environmental Protection Agency, the Occupational Safety and Health Administration, and the Army Corps of Engineers. The National Security Agency, the Central Intelligence Agency, and the Federal Bureau of Investigation hire large numbers of engineers annually, especially for designing and installing communication networks.

Again, it is important to think about one's long-term career while considering work in any area of the economy. Many engineers perform important work as consultants or business advisers, but they can only do so after gaining experience in manufacturing or government service. In the environmental field, for example, many specialists work first for the federal Environmental Protection Agency, or state-level agencies, and then go on to work for engineering firms that do clean-up projects. Another potential career choice is banking and financial services. Banks need to know as much as possible about the capabilities of the companies for whom they provide loans. Financial firms and stock investors may need advice on the potential of a new technology, the possible results of a shortage of materials or energy, or the effects of a new government regulation.

It is not easy to predict exactly what one would like to do ten or fifteen years from now. Don't be frustrated if you can't decide now. Make plans to give yourself as many options as you can for the future. If you believe that you might do well in research, plan to take extra science courses. If you think that you might want to be a business manager, take extra economics or business courses, and perhaps make preparations for attending business school after you finish your engineering education.

ENGINEERS AND COMPUTERS

Computers are something we hear about often; they have entered practically every aspect of business, research, education, and communications. There is a special relationship between computers and engineers. The goal of both is to handle large numbers of data and symbols rapidly and efficiently. Engineers also are the developers of most computer systems, especially the hardware. Today, one of the most dynamic elements of engineering is the effort to transform design processes or production methods into computer programs. The effort, when successful, helps engineers tremendously in improving the quality of finished products, and it can reduce the cost of engineering work.

Development of Computers

Ironically, computers weren't always useful to engineers. When computers first became widely available, the time it took to write a useful program, and the cost of the computer itself, made the effort uneconomical. Many engineers also believed that trusting a computer to perform a design function was risky. A small, insignificant mistake in a computer program could make an enormous difference in a final product. There was also the belief that as computers grew more powerful, they would put the engineers who programmed them out of work.

The track record over the past ten years or so has been quite different. As the level of programming experience grew, the programs got better and errors or difficulties were identified. The cost of computers has dropped dramatically. Engineers have found that the computer can be a valuable tool to extend their areas of responsibility, rather than be a threat to jobs. Today, the business of building computers specifically for engineers—a type of computer called the technical workstation—is one of the fastest-growing computer markets.

Current Trends

The use of computers is having profound implications for many traditional engineering tasks. Many mechanical engineers, for example, must design a part or component that will be cut from metal, or molded from plastic. In the past, this design was carried out by making drawings and then performing calculations on the various dimensions of the part. These calculations ensured that the part would hold up under use and that it would fulfill the purpose for which it was designed. When everything checked out, a final drawing was made (again, by hand), and the drawing was taken to the machinists who actually fabricated the part. Now, the design engineer draws the part on a computer screen, then loads a computer program that performs the checks automatically. Finally, another program translates the dimensions into instructions for an automatic cutting or molding machine, and the part can go into production rapidly.

In factory construction an enormously complex set of pipes, girders, vessels, and machines are put in place. Plant design engineers can now put drawings of all the components on a computer screen and then run an "interference check" to make sure that pipes don't cross each other, and that they make the proper connections.

In microelectronic-chip design, some circuits are so complex that they can be planned out only on a computer. Many circuit engineers are experimenting with programs called "silicon compilers" that will automatically create a circuit based on general guidelines that the engineer chooses. The compiler will then arrange the components in a configuration that minimizes the amount of space necessary. All these computer programs are continually being improved, and the confidence level of the engineers who use them is rising.

In the early 1980s, ABET made it a requirement for all engineering students to take at least one computer programming course. It would be wise to get additional training, especially through the use of computers in engineering courses. This is becoming easier as more schools acquire access to computing systems. There is no question that the computer will grow in importance in most aspects of engineering work in the future. Today's students should prepare for it.

SALARIES AND THE INTANGIBLE REWARDS

Here's the good news: at least initially, engineering graduates earn the highest pay of any college graduates. Surveys from such organizations as the College Placement Council show that engineers with B.S. degrees average around $34,000; with M.S. degrees, around $40,000; and with Ph.D. degrees, $50,000 and up. Over the long term, however, salaries tend to flatten out at a level that is still very high relative to all professionals, but not the highest.

Pay is a sensitive topic for most people. All other things being equal, nearly everyone would like to make more money for the work they do. The trade-off usually involves taking promotions that could require a shift in professional focus—going, for example, from production engineering to plant management, or design to sales and marketing. If higher pay is the sole object, an engineering degree certainly is no hindrance. Some of the highest-paid executives in American corporations started out as engineers.

Not all engineers put the highest priority on the biggest dollars. Many engineers, having gained the experience they believe they need, simply want to continue their current work for the remainder of their careers, getting better and better at that task. For many engineers this means that their pay will begin to level off, with each year's raise less than in preceding years.

Some companies, recognizing the value of such experienced engineers, set up a "dual track" for promotions through the ranks. Under this structure one can move through the ranks in the conventional manner, rising from design or production to management, marketing, or administration. However, one can also rise through another set of positions designed only for engineers and other technical specialists.

The highest pay, at least early in one's career, is with large corporations. The next highest is with small companies, and some of these are on a par with the salaries offered by government. Over the long term, salaries are higher for consulting engineers or engineers who start their own businesses. But with the higher pay go the uncertainties of the job market; one could lose a job if business conditions go sour or get less pay if sales or business goals aren't met.

It used to be true that most engineers, like workers of all types, were hired by a large corporation upon graduation and then spent the rest of their career with that employer, gradually rising through the ranks. Today, the commitment by employers to keep their staff through thick and thin is reduced. For their part most young workers today don't feel exceptionally loyal to their first employer either. The end result is that a certain amount of job-hopping is now the norm. This isn't necessarily bad; many engineers can gain valuable experience by dealing with a technical issue from different sides—for example, as a producer, a user, or a government regulator.

However, you must be ready to manage your career from the outset. Once you have joined a particular company in a particular industry, get familiar with all aspects of that industry. Read business magazines, talk with people in the field, and become knowledgeable about what trends are occurring. At the same time many engineers benefit by forming strong ties with their professional organization. Join the local section of your engineering society, go to the monthly meetings, and participate in the organization's annual events. The most valuable information about what is going on in an industry comes from talking with fellow engineers about it. Finally, don't rule out getting another job if that job will provide experience that you might value. And if you lose a job, be prepared to win another by having your resume and other information at the ready.

CHAPTER 3

CLASSIC ENGINEERING

The Industrial Revolution wasn't so long ago. If you have eld-erly grandparents, they may remember hearing for the first time about an airplane, or seeing the first of the Model T Fords. And when those grandparents were children, their grandparents could remember the first steam locomotives, the building of the first iron bridges, and the first electric light bulbs.

As the Industrial Revolution arose in Europe in the 1800s, the era of human or animal power transformed to an era of steam power, with engines that could cut wood, move merchandise, or thresh grain. Soon, a steady stream of artisans was coming from Europe to the United States and finding plenty of opportunities to exercise the art of manufacturing. As the century wore on, many new machines were invented in the United States. Soon, with one advance piling on top of another, a complex set of skills called en-gineering began to be devised. These new skills were taught at se-lected colleges and universities.

With one exception, it was during those times that the biggest, oldest engineering disciplines formed. The one exception was civil engineering, which used to be called military engineering. Military engineering was the practice of building the roads, bridges, and forts that armies on the move needed. The West Point Academy, founded in 1802, was the source of the first military en-gineers in the United States. Military engineering began to be taught in 1817. The Rennselaer Polytechnic Institute, founded in 1824, offered the first courses that were formally known as civil engineering.

Each of the engineering disciplines profiled here traces its roots back to the nineteenth century. Even so, an engineer of that era would hardly recognize anything going on in today's versions of these disciplines. Skill in drafting has given way to skill in manipulating images on a computer screen. Familiarity with steam locomotives has been replaced by an understanding of power supplies for communications satellites. Each of the disciplines listed here has a long history, but also has a new, up-to-the-minute look.

ELECTRICAL/ELECTRONICS ENGINEERING

To a large degree, saying that you are an engineer is synonymous with saying that you are an electrical/electronics engineer. The reason for this is the vast number of students studying electronics and the correspondingly large number of such engineers at work today. About one-third of all engineering students are in electrical/electronics programs. If computer engineering students and electrical engineering technology students are added in, the number approaches one-half.

Why is electrical/electronics engineering so popular and so dominant among working engineers? Most of the reasons can be found all around you—the radio you might be listening to, the television you might be watching, perhaps even the computer that you might be using. The twentieth century has been called many things, but perhaps the most accurate of all is to call it the Electric Age.

Spurred by new scientific discoveries, and then by practical inventions, electricity has flowed into every aspect of our lives. Electricity supplies light; power to run appliances and heavy machinery;

and communications such as the telephone, radio, and television. In much smaller quantities, electricity provides electronics—small devices that mimic large machines. And, of course, electronics is changing everything around us every day, through such pervasive devices as handheld calculators, computers, and control devices that help operate automobiles, airplanes, and homes.

There are roughly 439,000 electrical/electronics engineers at work today, according to federal data. Large as it is, the field is projected to grow even faster than all other engineering disciplines over the next decade, increasing by 40 percent to 615,000.

History

Electrical/electronics engineering has participated in some of the most momentous discoveries in science, both in terms of applying them to practical use and in uncovering phenomena that led to new scientific theories.

The science of electricity had only a rather quaint value as a curiosity for most of history. William Gilbert, an English scientist, characterized magnetism and static electricity around the year 1600, and Alexander Volta discovered that an electric current could be made to flow in 1800. In the mid-1800s, a variety of European scientists had established the general rules governing electricity, and, ultimately, theories involving electricity and magnetism were joined under a concept called electromagnetism (James Clerk Maxwell's discovery).

KEY INVENTIONS

Although most of the theory was developed in Europe, it was in America that most of the first practical applications appeared. These included the telegraph (Samuel Morse, 1838), the telephone (Alexander Graham Bell, 1876), the light bulb (Thomas Edison, 1878), and the electric motor (Nicholas Tesla, 1888). All these inventions—but most of all, Edison's light bulb—soon created the need for systems of generating and conveying electricity and manufacturing the telephones, motors, and bulbs that would use it. The American Institute of Electrical Engineers was formed in 1884, partly to professionalize the growing number of workers in the field and partly to prepare for international visitors expected at the International Electrical Exhibition being held in Philadelphia that year.

From that era to the 1930s, electrical engineering was primarily concerned with figuring out how to generate ever-larger amounts of power and refining the motors, transformers, lighting devices, and other machines that used the electricity. The needs of the electricity industries helped raise the standards of metalworking, machining, and general manufacturing because precisely shaped parts were essential to getting electricity to work correctly. Edison—who went on to create the phonograph (for both recording and playing back), the film camera, and a variety of electrical instruments—brought all his discoveries under the umbrella of the General Electric Company around the turn of the century. Bell's invention, the telephone, led to the creation of American Telephone and Telegraph.

THE BIRTH OF ELECTRONICS

The triode, a type of vacuum tube developed by Lee De Forest in 1907, and other inventions in England and the United States, led to a variety of devices that could employ a very weak electrical signal. This was the birth of electronics. These inventions helped make radio possible. They were quickly applied to telephony and then, in the 1920s, to commercial radio stations. Along the way, the Institute of Radio Engineers was established in 1912.

The distinction between electricity for power and electricity for communications (electronics) was the cause of some friction in the early years of radio. The overlap in technologies caused the two fields to encroach on each other's territory. Finally, in 1963, the two organizations united to form today's Institute of Electrical and Electronics Engineers, Inc. (IEEE). This division was even more strongly established in Great Britain and West Germany, where the fields were identified as "heavy current" (power) engineering and "light current" (electronics) engineering.

Over the years government sponsorship, especially from the military, helped spur the growth of electrical and electronics technology. The early days of radio were boosted by army interest in radio telephones and navy interest in shipboard communications. In World War II, radar and sonar were developed to improve battlefield conditions; these helped spur improvements in electronic components. Federal funding in that era also helped develop the computer. During the 1950s, intensive research was devoted to harnessing atomic energy to generate electricity first for ships and submarines and then for commercial application. And in the 1960s, the young field of space technology and missiles boosted the development of the integrated circuit.

MODERN TECHNOLOGY

The integrated circuit—sometimes called the microchip—is a dominant force in electronics technology today. Electronic devices through the 1950s needed a vacuum tube—pieces of metal inside a glass bulb. These tubes tended to be unreliable and to wear out quickly. In 1948 researchers at AT&T invented a solid electronic device, the transistor. This led to a widespread change in electronic designs, which were then called "solid-state." Then, in the early 1960s, researchers at Texas Instruments and Fairchild Semiconductor devised a way to build transistors on a tiny slice of silicon with small wires connecting them. Soon, a way to "write" these circuits on the chip with photographic techniques led to the dramatic situation we have today, with the number of electronic elements on a chip multiplying by the thousands each year.

The computer, primarily an invention of electrical engineers and mathematicians, has shared many of the benefits of the microchip. The basic, modern theories of computing were developed shortly after World War II. The first computers ran on vast arrays of vacuum tubes, with circuits being connected and switched manually. With the transistor, and then the integrated circuit, both the computing action of the computer and the storage of data (the computer memory) were greatly simplified and much less costly. The advent of the computer also led to the development of computer languages. Although it is still very possible to learn computer languages while studying electrical/electronics engineering, a more direct route is usually to study computer science.

The Current Scene

Electrical engineering affects a broad array of manufacturing systems, machines, communications networks, and transportation vehicles. It is hard to think of a machine or appliance without a microchip in it somewhere, and science fiction writers today have fun envisioning a time when we will have microchips implanted in our heads.

The oldest version of electrical engineering, the generation of power, is still a large field, but its size is dwarfed by the other specialties involving electronics. IEEE, organizes its membership according to these categories:

Division I: Circuits and Devices. This includes microchips, the larger circuits that microchips are wired into, lasers and electro-optics, and related solid-state devices.

Division II: Industrial Applications. This covers the manufacturing applications of electronics such as insulation devices, instrumentation and measurement, and electronic devices that control power.

Division III: Communications Technology. The fields included here are those most familiar to the general consumer: broadcast electronics, consumer electronics (radios, TV), communication devices (telephones, radio), and radios and similar devices in automobiles.

Division IV: Electromagnetics and Radiation. This represents the more advanced realms of communications such as those used for detecting aircraft. Subgroups include antennas and signal propagation, magnetics, microwaves, and nuclear and plasma sciences.

Division V and VIII: Computers. The Computer Society is the single largest division and includes computer hardware and data storage, networks, and electronics for everything from handheld calculators to supercomputers.

Division VI: Engineering and the Human Environment. Representing the outreach efforts of IEEE, this division includes engineering management, education, professional communication, and the social implications of technology.

Division VII: Power Engineering. These are the engineers at utility stations and those who design, construct, and maintain the generators and transmission systems.

Division IX: Signals and Applications. More types of advanced electronic transmission and detection are covered here, including acoustics, speech and signal processing, remote sensing, ultrasonics, and aerospace systems.

Division X: Systems and Control. Electronics are capable of controlling electrical and mechanical devices, even as electricity provides the power. Robotics, industrial automation systems, information theory, and engineering medicine are some of the subgroups of this division.

In 1992, when a survey of IEEE members was conducted, the proportions among these "primary areas of competence" (as the Institute put it) were as follows:

Table 1.   IEEE Membership Distribution

Circuits and Devices	12%
Industrial Applications	7%
Communications Technology	11%
Electromagnetics and Radiation	6%
Computers	24%
Engineering and the Human Environment	7%
Power Engineering	13%
Signals and Applications	6%
Systems and Control	8%
Other	5%

Except for the computer division, things are fairly evenly matched. But wait—there's more. IEEE isn't the only professional society managing the interests of electrical and electronics engineers. There is also the Instrument Society of America, which has about 40,000 members interested in industrial control and measurement. Not all of the ISA members are electrical/electronics engineers, and some ISA members are also IEEE members. However, the existence of ISA indicates the great number of engineers employed in the electrical/electronics field. There is a Society of Motion Picture and Television Engineers, 10,000 strong, and an equally sized Illuminating Engineering Society of North America. There are at least a half-dozen technical associations for computer specialists, with thousands of members (only some of whom have electrical/electronics backgrounds). Thus, the field of interest of electrical/electronics engineers is huge, and there are tens of thousands of them in most of the technical areas.

All engineering disciplines advance new technology, but for the past couple decades, no discipline has been moving faster than electrical/electronics engineering. This trend will continue for at least the next decade. Computers and semiconductors continue to evolve, and the technology for both of these refuses to "settle down" into something predictable and well-defined.

Job Titles

With so many avenues of technical development, the list of possible job titles for electrical/electronics engineers is lengthy. Some of the more common ones are as follows:

Circuit designer. Whether it is a microcircuit etched on a silicon chip or a circuit board on a piece of green plastic, these designers apply engineering principles to building circuits that will accomplish the desired objective. Circuit design is one of the most active areas for automated computer design. Many engineers have written computer programs to figure out the optimum arrangement of circuit elements.

Communications engineer. Most of the mass-market, long-distance communication networks—such as telephones, radio, television, and cable television—rely on these engineers to develop the best ways to send and receive the communications signal. Signal fidelity and immunity to electronic "noise" are constant goals.

Computer engineer. The design and construction of computers is one of the more prominent occupations of electrical engineers. These engineers must, of course, be familiar with circuits and microchips, but they must also have more than a passing knowledge of computer programming. For this reason many engineers interested in computer design major in that field (computer engineering), which offers more programming course work. (Employers appear to be evenhanded when choosing between the two specialties; the real distinction is the student's own course work and interests.)

Once a design has been completed and a prototype built, highly sophisticated tests are run to make sure that the computer can perform as expected. Usually problems are encountered that necessitate redesign. Similarly each computer coming off the production line is tested for reliability and performance.

Control engineer. The ability of computers and electronic devices to provide automatic control of appliances, machines, and manufacturing processes is generating high job demand for these specialists. One of the most dramatic possibilities is the use of artificial-intelligence computer programming to make processes "think."

Robotics engineer. Robotics suffered a downturn in business growth during the 1980s from which it is still recovering. But the long-term future is still bright. Robotics and control engineers share many of the same goals.

Power systems engineer. The design and operation of modern utility plants is extremely complex, more so when nuclear energy is involved. A widening gap between the capacity of newly built power plants and the demand for electricity is expected to generate high job growth for power engineers during the 1990s.

Of course, many other electrical/electronics engineers are involved in testing, maintenance, production, research, and other types of engineering work.

A Control Engineer

Jane has an exciting entry to professional work—an involvement in a first-of-its-kind artificial intelligence program to automate a common manufacturing process—water treatment. Just as water in an automobile's radiator helps keep the motor running, factories, power plants, and other facilities need to circulate cooling and heating water. This process water must be treated continuously to prevent corrosion and scale buildup inside the circulation pipes and pumps. Jane's company, a contract supplier of water treatment systems, wants to replace the constant attention to water conditions that requires several workers with a control computer that would inject the right mix of anticorrosion chemicals at the right time. It's heady work for an electrical engineer who just finished her master's degree.

One of the reasons Jane took this job was that there were few electrical engineers at the company and no one in her specialty—artificial intelligence. This gives Jane the chance to work on totally new systems; at the same time, it puts a spotlight on her that, if she were to fail, would be painfully bright.

The system Jane has designed marries a new semiconductor chip with an off-the-shelf personal computer. This PC is connected to standard industrial sensors and controllers that make chemical analyses of the process water and then open or shut the appropriate valves and start or stop pumps. The chip and PC have all the "intelligence" needed to compute appropriate control actions, but Jane needs to determine the exact goals of the system so that the right instructions can be programmed into the PC.

To do this Jane engages in what is called "expert systems development." She interviews several senior engineers who have either a chemical or mechanical engineering background. She finds out that the chemicals change the acidity of the water, causing dissolved salts to precipitate out before the process water runs through the system. She figures out the chemistry of the process and how valves and pumps are made to operate. Both of these areas send her back to her college textbooks to freshen up her understanding of chemistry and mechanics.

Finally, all the instructions are fed into the computer, and laboratory test runs show the system responds correctly. The system will now be field-tested at a customer's site, and Jane will monitor that work for most of the rest of the year to make sure the system works right.

Aerospace Electronics Engineer

The general term for electronics on aircraft is "avionics," and Raul likes to think of himself as an avionics engineer. He is part of a group of twelve engineers in a section of fifty avionics engineers, all of whom work for a major defense contractor. The fifty engineers are responsible for all the avionics for a new fighter jet the contractor is designing for the air force.

This project was unusual in aerospace industry practice because it involved competition between three defense contractors. Eighteen months ago, after submitting designs to the air force, each firm was given a sum of money, and had to invest money of its own, to build a prototype of the aircraft. The three prototypes were then tested competitively, and the winner was selected. Raul's firm won.

Now his team is involved in detailing the electronic subsystems that were only outlined during the competition. As is often the case, the electronic technology has changed drastically from the time when the original design was written until the new design must be prepared. In this case, silicon chips for a microwave receiver are being superseded by chips made of gallium arsenide. The subcontracting firm that is building the circuit boards wants to switch to the new chips, but they are more expensive. Raul must decide whether to go along with the subcontractor's wish, insist on the existing design, or propose something different.

After reading the new chip's performance criteria closely, Raul realizes that there is a trade-off: the new chip is more expensive, but it can perform better. In fact, using the new chip will simplify elements of another circuit that feeds signals into the board he is concerned with. He checks with the group responsible for this circuit and finds that they are willing to change their design to suit his.

After several discussions, they arrive at preliminary cost figures (more for the new chip, less for the other circuit board) and find that the costs roughly balance out. Because the gallium arsenide chip is more reliable, overall opinion tilts strongly in favor of making the switch. Raul now begins to prepare a report for his group leader. This report will become part of another report that the group leader will send to the company's upper management, and eventually to the Defense Department for review.

Education

Electrical/electronics engineering can be one of the most mathematical types of engineering. Whereas most other engineers are limited by the materials they use (concrete for bridges, steel for boilers), electrical/electronics engineers can work with circuits made of a great variety of materials, which can achieve a wide range of effects. Students of this field aren't required to take more math courses than most other engineers, but many of them do in order to improve their proficiency.

Another distinction of the electrical/electronics field is that many baccalaureate graduates go on for a master's degree—often earning it at night while working full-time during the day. The rapid pace of change in electronic technology makes it important to keep up, and one of the better ways of doing this is extending one's study. IEEE figures show that about one out of three electrical/electronics engineers earns a master's degree.

The typical courses for an undergraduate, beyond the normal requirements for all engineering students, follow two tracks: one for electrical and computer engineering and one for computer science/computer engineering. Course topics for electrical/computer engineers include:

- electromagnetic fields
- circuit design
- logic circuits
- computer architecture
- energy conversion

For the computer science/computer engineering major, the courses include more computer programming:

- computer hardware
- software engineering
- operating systems
- communications

A wide variety of technical electives exist in the many specialty areas of electrical/electronics engineering.

MECHANICAL ENGINEERING

Mechanical engineering is machines. Machines, power systems, factory production lines, computers, boilers, and pressure vessels are part of the mechanical engineering scene. The most obvious area of employment for mechanical engineers is the automotive industry, but fewer than 5 percent of all mechanical engineers work in that field. Companies that produce aircraft or electrical machinery, power utilities, and the federal government are other key employers.

Mechanical engineering is the second-largest engineering profession (behind electrical); there are 227,000 mechanical engineers at work today, and the projection is for above-average growth to 273,000 by the year 2005, according to federal data. Data from the College Placement Council show that the average offer for new B.S.M.E. graduates in 1994 was about $35,000. The average earnings for experienced engineers, in industries where mechanical engineers are employed, was around $57,000 in 1994, according to the Engineering Manpower Commission.

History

Mechanical engineering can trace its roots back to the very beginnings of the Industrial Revolution, from 1750 to 1800 in Europe and 1800 to 1850 in the United States. Perhaps the most important single invention prior to the actual creation of the profession was the steam engine, invented by James Watt in 1802. In short order this led to the steam locomotive and self-propelled boat. These two modes of transportation soon caused canals and railroad tracks to appear all over Europe and then in the United States. Somewhat later, the adaptation of the steam engine allowed the mechanization of agriculture to begin. It freed manufacturing plants from water power, which had been the traditional source of power for running conveyors and grinding and cutting machines.

The American Society of Mechanical Engineers was founded in 1888 by a group of leading businesspeople and the editors of a magazine called *American Machinist.* Two more developments created new demand for mechanical engineers around that time. One was the automobile, powered by the combustion of oil or other fuel. The other was the application of electricity for lighting—one of the many discoveries that came out of Thomas Edison's laboratories in the 1800s. With an electrical lighting system for the home or for city streets came the need for power generators and electrical conveying equipment. And with the automobile, the need for precisely machined metal became critical, as well as a more formalized method of assembling the components into finished autos.

TWENTIETH-CENTURY DEVELOPMENTS

By the 1920s, with the addition of the airplane to the growing number of methods of transportation, social commentators were hailing the establishment of a "machine civilization." The excitement of that time matches the excitement being generated today by computers.

Also by this time, the central element of mechanical engineering began to be clarified: the generation and use of power. Power means automotive horsepower, the watt-hours an electric utility generates, the thermal units that a heating system produces, and the thrust of a space rocket. Taking some form of energy and converting it into useful work is the key activity of mechanical engineers.

A good example of this idea and of how mechanical engineers work can be found in a unique publication sponsored by the American Society of Mechanical Engineers. That publication is the *ASME Boiler and Pressure Vessel Code.* The 1989 edition (the 75th) runs 11,000 pages! Luckily, it is now available on a single computer-compatible CD-ROM disk. The *Boiler and Pressure Vessel Code* provides all the details on how to design, assemble, and test a tank, usually made of steel, that contains heated water or high-pressure steam. The rules of the *Code* are written into state and national safety standards, and because they are followed with great exactitude by mechanical engineers, there are relatively few boiler explosions today. These explosions were a common occurrence in the 1800s on steamships, railroad trains, and in heaters and power systems.

During World War II the concepts of mechanical engineering were very important to the design of aircraft, tanks, and ships. This period also saw the first joining together of mechanical devices with electronics. In fact, the first computers, which were very mechanical devices, were developed by military-funded projects aimed at developing a way to compute the firing trajectories of artillery shells. World War II also brought about the development of the space rocket and the jet engine, which are commonplace today.

The mating of electronics and computers also made automatic machines—robots—possible. This field, expected to become a fast-growing, gigantic business a few years ago, is quieter right now. But most forecasters expect that as the cost and difficulty of guiding machines by computer become reduced, robotics will become a key technology in the future.

The Current Scene

As one of the largest engineering professions, with employment opportunities across most types of manufacturing, the mechanical engineering profession is subject to the same ups and downs as the United States economy as a whole. The economy has been performing well for most of the past decade, and the forecast is for continued, if slightly slower, growth in the future.

In the 1970s, as energy prices skyrocketed, the mechanical engineering profession became one of the most prominent in the effort to conserve energy and make energy-intensive processes and machines more efficient. This need to conserve energy led to minimum mile-per-gallon ratings for automobiles and energy-efficiency ratings on household appliances, among other applications. A tremendous amount of work was performed by mechanical engineers in redesigning all these machines.

Today, with energy prices mostly stable, energy efficiency is a less-critical factor. What has come to the fore now is the need for less pollution. Exhaust gases from cars, aircraft, power plants, and heating systems have caused air pollution, acid rain, and the decrease in the ozone layer of the earth's atmosphere (which protects us from harmful solar radiation). The most profound problem, however, is the warming of the planet, caused by an increase in the amount of carbon dioxide gas in the atmosphere. Although still subject to much research and debate, a number of scientific analyses point to this gas as the cause of the "greenhouse effect" in which heat and radiation enter the earth (mostly from the sun), but cannot escape through the thickening blanket of carbon dioxide in the atmosphere.

The global warming issue is troublesome because when something is burned or ignited, carbon dioxide gas is almost always generated. Striking a match, lighting a wood fire, or even breathing produces carbon dioxide. To reduce the amount of combustion in the civilized world will require enormous changes in how we do things. Mechanical engineers will be at the forefront of contending with this issue in the future.

New machines are appearing all the time, including more advanced robots. In the past, a mechanical engineer used to sit at a drafting table figuring out the dimensions of the parts of a piece of equipment. Then a machine shop cut and ground the metal to form prototypes of the parts. Today much of this design work is done on a computer with specialized programs called CAD (computer-aided design). The vision of the future—which mechanical engineers are gradually turning into a reality—is to develop designs on a computer, test them with other programs, then send the designs to automated production machinery that will fabricate and assemble the parts. This procedure will make possible very fast redesigns to meet customer demands and lower production costs.

To give an idea of the many types of work mechanical engineers do, take a look at the following list. These groups are the divisions or special interest areas of the members of the American Society of Mechanical Engineers.

- *Basic engineering,* including fluids, applied mechanics, heat transfer, tribology (the study of lubrication), and bioengineering
- *General engineering,* covering management, safety, and technology and society
- *Manufacturing,* involving materials handling, production engineering, textile engineering, process industries, and plant engineering and maintenance
- *Energy conversion,* including fuels and combustion technologies, internal combustion engines, power and nuclear engineering
- *Materials and structures,* comprising materials, pressure vessels and piping, nondestructive evaluation engineering (i.e., testing materials without destroying them), offshore mechanics, and arctic engineering

- *Energy resources,* involving petroleum, solar energy, ocean engineering, and advanced energy systems
- *Environment and transportation,* covering the topics of rail, aerospace, environmental control, solid-waste processing, noise control, and acoustics
- *Systems and design,* comprising dynamic systems and control, design engineering, computers in engineering, electrical and electronic packaging, and fluid power systems and technology

This list also gives one a sense of the job titles available, ranging from design to production, testing, or computer analysis. Mechanical engineers can spend all their time writing computer software or work full-time on a factory floor. Many mechanical engineers also ascend to corporate management. The range of opportunities is very broad.

Job Titles

Wherever there are machines, there are mechanical engineers. Of course, a higher concentration of engineers are in the areas of industry where machines are produced—automotive, aircraft, machine tools, and power generation systems. Some of the typical job titles are:

Design engineer. These engineers work with computer programs, laboratory models, and prototypes to develop new machinery or components. Today many types of machines are developed with electronic components. The electronics provide control and measurement capability, and the mechanical devices transform the electronic instructions into physical action. Thus, a knowledge of electronics and control theory is a help.

Manufacturing/production engineer. The 1980s were characterized by an enormous adjustment to robotics and computer controls in manufacturing processes. Higher quality, faster throughput—all at lower costs—remain the challenge. Mechanical engineers with this kind of experience can move rapidly into corporate management.

Maintenance engineer. Keeping production lines, power plants, and other machinery running smoothly is an essential part of manufacturing. Maintenance engineers work with mechanics and technicians to get a balky machine back into operation; the difference is that the mechanical engineer also looks at *why* a machine is failing. Is it too old and needs replacement? Is there a design flaw? Is it being used incorrectly on the production line? Answering these questions requires sophisticated analysis of the problem.

Power engineer. Some power engineers have an electrical engineering background, but most of the others have a mechanical engineering degree. In many applications, including electrical utilities, generating stations, and aircraft companies, the actual power-production machine is a turbine generator or an internal-combustion engine. Mechanical engineers have a strong understanding of both.

Automotive engineer. As with the power engineer, many, but not all, automotive engineers have a mechanical degree. They can also be electrical, chemical, or industrial engineers, among others. Automotive engineers have their own professional organization—the Society of Automotive Engineers. (See Appendix B for details.) They work in design, production, and testing for automakers and their suppliers.

Reliability and testing engineer. Long-term performance is a key characteristic of well-built machines. These engineers develop testing methods and review processes to determine how well equipment stands up to use.

A City Planner

Shirley works for an urban planning/technology group that does contract work for governments. Her firm has a contract with one of the California air-quality districts that oversees environmental issues on a regional basis within California. Shirley's firm will help develop a future air-pollution-reduction plan for the district. The issue she has to decide today is how to apportion essential services, which may involve pollution-generating machines or systems, while reducing overall pollution levels.

The region her firm is examining has a mix of power companies, agriculture, light and heavy manufacturing, and mass transit systems. Of course, there are also many people with homes that have heaters and cars that have exhaust pipes. Conceivably, she could write a proposal to ban all private ownership of automobiles and let industry grow much larger. She could also mandate the shutdown of all heavy industry and let people enjoy as many cars as they want.

Obviously, both these cases are extremes that could not be put into effect. Some set of conditions between the two must be established. Shirley applies her knowledge of trends in the technology of power generation and automotive technology to the issue of pollution generation. She knows that if a number of programs for low-polluting alternative fuels are established, automotive pollution could be reduced. She begins checking with other engineers in the transportation business to find out about other options.

The contract will go on for several months, and after that, public debate over the various options will begin. Shirley knows that this knotty issue won't be solved any time soon, but she also knows that facing it is inevitable. Something will have to be done about air pollution in this region, and the planning had better start now.

Production Engineer

Aaron is a plant engineer at a paper mill, but at this time he considers himself to be a high-class mechanic. The machine he is running is an enormous wood chipper—a machine that has rotating shafts powered by electric motors. It cuts logs of Georgia pine into small chips, which are then treated with steam and chemicals to produce the pulp out of which paper can be made. This chipping machine is a new addition to the plant; the previous chipper was retired after a decade of service. For the past three months, Aaron has been monitoring the machine's performance on an hourly basis.

He takes a personal interest in the machine because he had a say in deciding what vendor would supply the machine and in writing the specifications upon which it was assembled. A new feature of the machine is the use of what are called adjustable-frequency drives. This technology, being adopted by many different industries in many applications, has only recently been examined for wood chippers. Aaron argued strongly for its use because it offers the potential of reducing electricity consumption in the chipper. And electricity prices, while relatively low in some parts of the country, have been rising steadily. The machine will earn its cost several times over if the power consumption drops as planned.

The problem right now, though, is that power consumption is still relatively high. Aaron has been taking measurements of the "load" (the amount of force being applied) to each of the motors that drive the rotating shafts. He thinks he has located the problem. Each motor is supposed to run independently, adjusting its output as logs of varying lengths and shapes enter the chipping section. But because of the way the power controls on the motors have been specified, one motor controller is dictating the speed at which all the other motors are running. Some of them are thus running faster than necessary.

To prove this theory, Aaron uses a downtime period to rewrite the software that runs the controllers. This software isn't a computer program. Rather, it is simplified instructions that are used by a device called a programmable-logic controller (PLC), which in turn manages the frequency controllers on each motor. Aaron can do this rewriting at a personal computer in his office, debugging the instructions, and then transfer them to the PLC on the chipper. Aaron does so and finds that power consumption is dropping nicely once the chipper begins operating again. The situation will still have to be monitored, however.

Education

Mechanical engineers take the same chemistry, physics, and math courses that most other engineers are required to have during their freshman and sophomore years. In addition there are specialized courses for mechanical engineering, including:

- statics, dynamics, and kinematics—the study of how motion is propagated by structures
- control theory—how mechanical motion can be started or stopped by control devices
- thermodynamics
- mechanical design
- computer systems
- metallurgy

About one out of three ASME members has a master's degree (M.B.A. and/or engineering).

CIVIL ENGINEERING

To build may be a primal urge. Our constructions, while they may be simply for shelter or transportation, often include aesthetic touches that are there to make us feel good about what we have built. Thus, bridges have geometrical designs intended to support weight, but they also have an artistic detailing and a "look" that defines the era in which they were built.

In constructing buildings, highways, and bridges, civil engineers work with architects to develop the appearance of the structure. Ugly buildings represent a failed communication between the two professionals; a building that falls down, or cannot be maintained, represents an equivalent failure, but one that the civil engineer should have prevented.

But civil engineering is much more than erecting skyscrapers or bridges. Civil engineers are trained in the interactions among structures, the earth, and water, with applications ranging from highways to dams and water reservoirs. Deeply involved with specifying appropriate construction materials, many civil engineers are also employed by the manufacturers of those materials. And, since constructing a large building or public-works project can involve elaborate planning, civil engineers can be outstanding project managers. They sometimes oversee thousands of workers and develop advanced computerization and planning policies.

Most significantly, many civil engineers are involved with preserving, protecting, or restoring the environment. Most water treatment and water purification projects are designed and constructed by civil engineers (in these two areas, many of them are known as sanitary engineers). A growing number of civil engineers are involved in billion-dollar projects to clean up toxic industrial or municipal wastes at abandoned dump sites. Civil engineers engage in such diverse projects as preserving wetlands or beaches, maintaining national forest parks, and restoring the land around mines, oil wells, or factories.

There are about 173,000 civil engineers at work today, according to federal data. This total is expected to rise by approximately 24 percent—slightly above the average for all professions—by the year 2000.

History

Construction is one of humanity's earliest organized activities. Therefore, it is no accident that civil engineering was one of the very first to be formally set up (in the early 1700s in France). In the United States, the American Society of Civil Engineers was organized in 1852—the first national engineering society in the country.

In the mid-1800s, and through to this day, one of the central tasks of civil engineers was the design of roads and bridges. The history of American technology can be traced in the bridges around the country, with wood being replaced by iron and steel. Then those beams or girders were replaced by steel cables in such landmark structures as the Brooklyn Bridge (completed in 1883). In this century, new forms of concrete and steel-reinforced concrete are the most common bridge-building material. The advent of the automobile set off an avalanche of highway construction, culminating in the legislation that set up a national highway trust fund in the 1950s. Over the past forty years, thousands of miles of interstate highways have been built, redefining the landscape of America and its cities.

The "civil" in civil engineering refers to the discipline's involvement in public works, including government buildings, military bases, water treatment works, mass transit systems, airports, shipping ports, and parks. Because of this involvement, many civil engineers find themselves employees of, or suppliers for, local government. This relationship, combined with the requirements for public safety, translates into a high degree of professionalism. Civil engineers with professional engineer (P.E.) licenses are fairly common, and if a civil engineer expects to perform publicly funded work, getting the P.E. license should be a priority.

The computer has had a significant impact on the civil engineering profession. The traditional stereotype of a civil engineer is one who is carrying around a large set of blueprints of the details of a structure. The blueprints had been generated by large staffs of draftspeople working with precise mechanical pencils and rulers. Today most of that work is done on high-powered, graphics-rich computer workstations. As the designer adds components, the computer keeps track of their location and can generate accurate drawings from any perspective. "Bills of materials"—the list of types and quantities of construction materials—are added up automatically. The most advanced workstations allow the viewer to "walk through" a computer-generated animation of what the yet-to-be-built structure will look like.

A key word that arose in the 1980s and will remain important for civil engineers for many years to come is "infrastructure." This term refers to the facilities that local, state, and federal governments provide in order for private industry to expand, or for improving the services for private citizens.

The Current Scene

Construction is a key part of the overall American economy. Data from the U.S. Department of Commerce show that more than $400 billion are spent each year on new construction and about another $100 billion on repair and maintenance of existing structures. To this half-trillion-dollar total can be added the $50 billion or so that is paid for construction materials. Many civil engineers specialize in the development and production of new construction materials.

Infrastructure demands will remain a key part of the civil engineering scene for years to come. With most of the interstate highway system in place, there is now a need for maintaining it and for adapting it to new traffic patterns. Similarly, the nation's airports, railroads, and waterways need regular refurbishing. The housing stock of private homes, apartment buildings, and facilities like colleges or hospitals get renewed on a steady basis.

Civil engineering also comes to the fore when social changes foster new development. In the 1950s and 1960s, much business growth was creatd by the construction of the interstate highway system. In the 1970s the prominence of the Sunbelt became apparent; northern states have been losing population, while southern and western states have gained dramatically. Such population swings require new construction for roads, schools, water systems, and housing.

Overall, however, the civil engineering field in the United States is not as dynamic as it was two or three decades ago when the interstate highways were being constructed, when new communities were popping up all over the land, and when public funds were more available. The United States economy was also growing at a faster clip during the 1950s and 1960s, resulting in a higher demand for new factories. Today, the reduced demand for civil engineers can be seen in the slightly lower salaries that civil engineers earn coming out of school. Most salary surveys indicate that B.S.C.E.s get 10–20 percent less than other engineering majors. It is still, however, a very healthy salary.

This is not to say that one cannot have a wildly successful career—and make lots of money—in civil engineering. Perhaps more than in most engineering professions, civil engineers work as partners in privately held firms. These firms are set up the same way a law firm is, with several senior partners sharing the profits and junior partners and associates earning salaries until they move up to senior status.

The business is what you can make of it. At the same time, saying that there is less growth in the American economy is not the same as saying there is no growth. New factories are being built, new skyscrapers and bridges are going up across the land, and more environmental work is being scheduled.

Job Titles

Over the past couple of decades, the broad field of civil engineering has become specialized in a number of areas. Civil engineers with one type of experience are able to shift to another area, but the real career growth occurs as one becomes an expert in one of these specialties:

Structural engineer. This is the classic civil engineer, concerned with designing walls, towers, bridge spans, dams, or foundations. A knowledge of construction materials and methods is combined with analytical techniques that determine how much weight or mass a structure is carrying, what forces it must withstand (such as wind or water), and, in cases where an architect is involved, how best to accomplish the architect's vision.

Construction engineer. This engineer works at the construction site transforming blueprints and drawings into cement and steel reality. Besides understanding the principles by which a structure was designed, the construction engineer must manage the actual work. This can involve elaborate scheduling and planning so that materials and workers are brought to the site and complete their purpose in the proper order. Time pressure and an awareness of the financial elements of a project are constant objectives. Because the work is done outdoors, sometimes in very remote areas, one must be prepared for a lifestyle of "camping out" in temporary quarters for long stretches of time.

Surveying and mapping engineer. Even before a design is worked out, and as construction begins, teams of surveying and mapping engineers are at work. They use electronic instruments and even satellites (which provide detailed overhead views) to measure the dimensions of the project. Some construction projects can cover dozens of square miles of territory. Elevations must be determined and calculations made regarding how much earth needs to be moved.

Transportation engineer. Do you prefer to travel by plane, train, auto, or bus? Transportation engineering has provided th wealth of traveling options we enjoy today. Highway design is constantly being improved by making roads safer and, in urban areas, making plans for handling increased traffic. Transportation engineers also oversee the design and construction of mass transit systems such as subways, which require tunneling, railway construction, and research on commuting plans.

A subspecialty within transportation engineering is the pipeline engineer, who determines the movement of water, oil, or gas through pipelines. In certain aspects this field is comparable to highway design, with the distinction that a liquid is being conveyed, rather than vehicles.

Environmental (sanitary) engineer. These engineers specialize in water and wastewater projects, land remediation, aqueducts, and garbage disposal. This field is currently one of the fastest-growing of all engineering specialties; billions of dollars are being allocated for water and wastewater treatment, and for methods of processing garbage and other solid wastes.

Hydraulic and irrigation engineer. Utility companies and many factories, farms, and river or lake barges depend on a steady source of water. These engineers perform the planning, design, construction, and maintenance to keep those water supplies available. Dam design and construction, flood control, and the design and construction of reservoirs, wells, and aqueducts are all common projects. It used to be that hydraulic engineers were concerned with draining swamps and straightening waterways. These days, they are as likely to be constructing swamps and estuaries to preserve the environment and provide reserves for fish and wildlife.

Geotechnical engineer. Along with geological engineers, these engineers help determine the underlying rock strata that affect roadways, water reservoirs, bridges, and other large structures. Earthquake planning and preparation also fall into this category.

A Surveying Engineer

John is feeling more relaxed these days, realizing that a major contract his new firm undertook is now winding down. As one of the founding partners in the firm, he has had several anxious months in the past couple of years as the firm waited for business to come in. But now, on the verge of completing its biggest project to date, he can breathe a little easier.

John's firm specializes in site preparation work. They do surveys to determine the size and shape of the terrain at a project site, as well as underground probes to find where bedrock and underground streams might lie. Together, these tasks must be performed in order to figure out where bulldozers must cut, or where explosives must be used to move rock strata.

A new hospital is being built, over several hundred acres, in a rather remote corner at the end of a valley; the remoteness is expected to add to the restful feeling for patients and visitors to the hospital. The general contractor—the firm that will be responsible for the overall project—gave John's firm a contract to do the site preparation. John was the overall supervisor for his firm's work.

The first step was a survey, which involved carrying electronic surveying equipment up and down the hills on the site. John also hired a photogrammetry expert to take aerial photographs and to convert those to dimensional drawings of the site elevations. John spent a good part of the previous spring hiking with the surveying crew, but his main responsibility was to monitor their progress, not to do the actual surveying.

The next step was to bring a well-drilling crew out to the site, not to dig wells, but to take core samples of the ground. John monitored this work closely, having specified where the crew was to dig. He also had a hand in analyzing the core samples as they were delivered from the site. The workers found a couple of underground streams—not unexpected in terrain such as this—and one big surprise. A low-lying area in one corner of the site had apparently been used before as a dump by a company that burned coal. The workers found samples of ash that were identifiable as coal wastes, as well as a heavy concentration of coal tars. The general contractor, and the hospital management firm, were extremely interested in these results because new federal and state laws dictate that the landowner must clean up any harmful waste products found at a site.

John's firm helped characterize the size and composition of the waste materials, which were under a dozen feet of soil. Chemical laboratory tests showed what types of waste products were at the site, and John was able to verify that the wastes would have to be removed.

As John began finishing up the final report of site preparation survey, he was proud of his firm's professionalism. They handled all the details of the surveying, provided the guidelines for where earth had to be moved, and saved the owner from a big problem by finding the dump site before construction began. John expects that this project will lead to more contracts with other general contractors in the future.

Education

In addition to the core courses that nearly all engineering students attend, civil engineers choose from an extensive list of civil engineering classes. Some students make selections based on the specialty they desire to follow; others, not having any specialty in mind, try to fit in as many of the civil engineering courses as possible. The list of civil engineering courses includes:

- surveying and design graphics
- materials design and specification
- geology and hydraulics
- structural analysis
- soil mechanics

- sanitation engineering
- transportation engineering
- geology
- environmental engineering
- oceanography
- steel and/or concrete structures

About a third of the students earning a B.S. degree go on to take a master's, in which specialization in one of the civil engineering programs is intensified. In addition, a proportionately higher number of civil engineering students take the time to qualify for a professional engineer's (P.E.) license. This license is often a requirement for being involved in public works or for buildings, so many civil engineers need the license in order to practice.

CHEMICAL ENGINEERING

Chemical engineering takes the knowledge that chemists obtain in laboratories and tries to turn it into tonnage quantities of materials for social needs. Not all the materials are provided by the ton, or even by the pound. The latest biocompounds coming out of genetic engineering laboratories are extracted and purified via chemical engineering technology; some of them are worth more than $25,000 an ounce.

There are about 52,000 chemical engineers at work today. The Bureau of Labor Statistics projects that that number will increase by 19 percent, to about 62,000, by the year 2005. Starting salaries for B.S.Ch.E.s was just under $33,000 in 1989, according to the College Placement Council. The median salary (with 50 percent of engineers above this number and 50 percent below) for experienced chemical engineers in 1994 was $56,300, according to a survey conducted by the American Institute of Chemical Engineers. Engineers with advanced degrees (master's and others) earn $10,000–$15,000 more.

History

During the period 1850–1900, a variety of industries grew in importance in the United States. The new products included paper, fertilizers, refined metals (as the mineral-rich American West opened up), and energy products including coal and, later, petroleum. The industry where the most chemistry knowledge was applied was in textile dyeing, and most of the technology for this field came from Germany.

Industry got by with a combination of chemists (often called "industrial" or "applied" chemists) and mechanical engineers. The chemists devised the reactions that produced valuable products, and the mechanical engineers devised the vessels and equipment that carried out these reactions.

RISING PROFESSIONALISM

By the turn of the century, the needs of industry for mechanical engineers with specific training in chemical processes led to the establishment of a number of chemical engineering programs. The very first, by most accounts, was at the Massachusetts Institute of Technology in 1888. There was some resistance to the establishment of these programs by chemists' professional organizations, but by and large the needs of industry overcame this resistance.

In 1908 there were a sufficient number of well-established chemical engineers that they could consider starting an organization of their own. The American Institute of Chemical Engineers came into being in that year in Philadelphia.

The profession grew gradually during the 1900–1920 period, until the end of World War I. One of the outcomes of that war was the embargo of many critical materials from Germany, which necessitated a scramble in the United States to set up production of these materials. Many German-owned companies were also expropriated as a means of reparation for the cost of the war. These companies needed United States citizens for management.

UNIT OPERATIONS

After World War I, the formal technology of chemical engineering became established, centering on a concept that is important for chemical engineers to this day. This concept is called *unit operations*. It was devised by a chemical engineer, Arthur D. Little, who started a consulting company that still exists. A unit operation is simply a piece of equipment, such as a tank, distillation column, or heater with a certain amount of raw material or intermediate product passing through it. A chemical engineer analyzes this unit by calculating how much material and energy is in it; what the piece of equipment is doing to change these amounts; and what, if any, chemical reactions are going on inside the equipment.

For example, a common unit operation in mineral processing is *calcining*. Dustlike particles of a mineral are dropped into a heated chamber, and hot air or a fire is introduced. The process causes the dust to bind together and to be purified of light contaminants, such as water or carbon dioxide. The chemical engineer asks these questions: How much heat do I need to apply to get a certain quantity of calcined material? How complete is the process of driving off contaminants? Is it possible to destroy the chemical products that I want in the final step, and if so, how can I prevent it?

Unit operations came into being as a concept because it enabled chemical engineers to design large-scale, continuous processes to produce larger amounts of product faster and more efficiently. Previously most chemical processes involved dumping something into a vat; heating, mixing, or reacting it; and then opening the vat and shoveling or pouring the result out. Product quality and production cost varied from batch to batch. Unit operation theory also unified processes that appear to be different on the outside, but in reality are the same.

Unit operations also provide a way to conceptualize chemical factories. From the outside, they look simply like a forest of vessels, pipes, smokestacks, and conveyor belts. But from a design engineer's perspective, the factory is simply a long series of unit operations strung together. Pipes or conveyor belts carry the material being processed from one unit to the next. It is successively purified, treated, or reacted and then formed into the desired end product.

The theory of unit operations came along just in time because around 1920 the United States was going through another energy crisis caused by the booming demand for automobiles and gasoline. Petroleum refiners needed to increase capacity and get more gasoline out of each barrel of crude petroleum they received.

CHEMICALS AND PLASTICS

Chemical engineering got another boost during the late 1930s, and especially during World War II. Again, critical materials were cut off—in this case, natural rubber from the Far East. Around the same time, chemical researchers at Du Pont Company invented nylon. At the time nylon was looked on as a type of artificial silk. It led to a literal flood of new materials, now know as polymer plastics: polyethylene (the plastic in shopping bags), polycarbonates ("plastic glass"), polyester, acrylics, styrenes, vinyls, phenolics—the list goes on and on.

The commercialization of these polymers helped boost the chemical industry into one of the key parts of the American economy, with annual revenues in the hundreds of billions of dollars. Most of these materials, as well as many chemical liquids or gases in common use today, are derived from crude petroleum. Thus, the oil industry and the chemicals and plastics industries are inter-twined (although a far larger volume of petroleum goes to the making of fuel than toward plastics and chemicals).

ENERGY AND THE ENVIRONMENT

During the 1950s and 1960s, chemical engineering helped de-velop nuclear energy, although in later years the field of nuclear engineering came into its own. With a knowledge of how petro-leum can be processed, chemical engineering crosses over into many aspects of the power utility field. Chemical engineers have also been involved in such areas as solar energy, energy conserva-tion, and coal processing.

Also during the 1950s and 1960s, America began an intensified effort to reconcile the problems of heavy industrial production with the environment. These efforts culminated with the creation of the United States Environmental Protection Agency in 1970. Chemical engineers were in demand then to help devise cleaner production techniques and to clean up past pollution; the demand for such professionals is rising rapidly now.

The demand for chemical engineers reached all-time highs in the late 1970s as the nation prepared for an era of energy indepen-dence by developing its own synthetic fuels industry, which was to have used such domestic sources of energy as coal or oil shale to reduce the need for imported oil. But the price of oil nose-dived in the early 1980s, the synthetic fuels industry never really got started, and thousands of chemical engineers were out of work.

By the mid-1990s chemical engineering had returned to a more traditional proportion of workers in various industries. Roughly half work in chemical production (or firms that serve this indus-try), and the remainder are spread throughout the industries that process materials and energy. About 9 percent work for environ-mental service organizations, according to a survey by the Ameri-can Institute of Chemical Engineers.

The Current Scene

Most chemical engineers identify their field of work as the chemical process industries (CPI). This category cuts across many types of manufacturing and services. The CPI includes:

- chemicals, including petrochemicals
- fertilizers
- pulp and papermaking
- rubber and plastics
- petroleum refining
- pharmaceuticals
- processed foods
- stone, clay, glass, and ceramics
- energy and fuels
- paints and specialty chemicals
- metals and minerals refining
- engineering design and construction
- environmental services

Most of these industries are traditionally considered "smoke-stack" industries, dealing with taking raw materials from the earth and transforming them into the products we need. For a while in the early 1980s, there was talk of a decline in the smokestack industries in the United States, with production moving abroad and the economy becoming more service oriented. But people soon realized that America cannot sustain itself without industry; someone has to obtain the essential materials that are the basis of all our finished goods. National wealth is created when something of no value in the ground is transformed to something of high value that people can use.

The American Institute of Chemical Engineers, together with the National Research Council, looked at the research needs of chemical engineering technology in 1988. Their report, "Frontiers in Chemical Engineering," was specifically geared toward highlighting research topics that deserved greater attention from engineers, industry, and the federal government. But it can conveniently be looked at as an indicator of where the chemical engineering profession is headed. New research performed today will create jobs for engineers tomorrow. A few chemical engineers are already involved in the newer areas; the more mature ones will see new growth in the future.

The seven areas are as follows:

1. *Biochemical and Biomedical Engineering.* Because of their understanding of fluids and chemical reactions, chemical engineers have insight into many processes going on in the human body or in forms of life such as animals or bacteria. Today's biotechnology revolution in pharmaceuticals and agriculture, which includes genetic engineering, depends on chemical engineers who extract and purify new drugs.

2. *Electronic, Photonic, and Recording Materials and Devices.* This covers the full range of consumer and industrial electronics including microchips, recording tape or compact disks, fiber optics, and printed circuit boards. In addition to being finely designed structures, these objects are physical things that must be manufactured to precise shapes and chemical structures. Microchips, for example, are melted silicon that is then crystallized, cut, etched, cleaned, and coated—all chemical processing steps.

3. *Advanced Materials.* Yesterday's wonder material, like nylon, is today's commodity. But today's wonder materials are still being developed. These include ceramics that bounce, metals that don't rust, superconductive magnets, and plastics tougher than steel.

4. *Energy and Natural Resources Processing.* The mining and minerals industries, and petroleum production, are quieter today than they were in earlier years, but they have by no means disappeared. New technology is making more gold extraction possible and permitting the recovery of natural gas from uneconomical wells. This field employs a lot of

chemical engineers now and could employ many more in the future.

5. *Environmental Protection, Safety, and Hazardous Materials.* It is not a coincidence that chemical engineers are experts in dealing with environmental and hazardous material problems, since many of these materials are manufactured by the CPI. Chemical engineers are being employed by the CPI companies to help minimize waste production. Government agencies, like the Environmental Protection Agency, also hire chemical engineers to regulate polluters.

6. *Computer-Assisted Process and Control Engineering.* Modern chemical plants or power utilities are amazingly complex machines, but many of them can be run by a handful of people sitting in a control room. Many chemical plants and power utilities are fully automated. This technology, coming from the computer and electronics industries and adapted to real-world needs, is an especially popular job area today.

7. *Surface and Interfacial Engineering.* An enormous amount of new technology involves how materials are coated or treated. Many new jobs will be created in this field. For example, it is now possible to extend the life of an artificial hip implant by firing metal atoms into its surface with a machine called an ion implanter. The precise depth and concentration of the atoms can be controlled. Most of the surface engineering work is going on in the electronics industry, metals processing, and production of polymers.

Job Titles

Because chemical engineering covers a broad spectrum of industries, there are many different types of jobs to obtain. Some of the more common ones include the following:

Plant engineer. These engineers are responsible for keeping a chemical process running. They work with machinery operators, monitor product quality, and troubleshoot problems as they occur.

Project engineer. These engineers adapt new technology to existing processes or help build and run new process units. Most chemical process plants are in a continual mode of upgrading or modernizing. Project engineers must know both how a process works and how it could be improved.

Design engineer. When new processes must be created, a design engineer will calculate the sizes and types of equipment appropriate for the process. The pilot plant—a small unit used to calculate full-size designs—is an essential tool for these engineers. Many design engineers work at engineering/construction companies, putting together the building specifications for new plants.

Researcher. Like all engineering disciplines, there are opportunities to perform pure research in chemical processing. Research chemical engineers may work with chemists and physicists, running experiments in a laboratory, or with other chemical engineers in scaling up a process to commercial size.

Environmental and workplace-safety engineer. Chemical engineers can do a lot to prevent pollution and to protect the well-being of plant workers. These engineers help design processes to minimize emissions or reduce waste generation.

Control engineer. One of the dominant trends in chemical engineering today is the adaptation of computers and automatic controls to manufacturing processes. Control engineers help select and install electronic hardware; they also write or oversee the computer software to run the plant.

A Pilot Plant Engineer

Tom is starting the new year with a specific goal in mind: developing the equipment to test a new processing route for purifying water. Pure water is valuable because it can be reused in the production processes of the chemical plant where Tom works. The plant is a huge complex where hundreds of different compounds are manufactured including natural gas, petroleum distillates, and metals. Purified water is also important because if the water can't be reused, it must be discharged to a local river, and strict environmental regulations must be met.

Wastewater from the chemical complex comes to a centralized treatment facility where it is cleaned by an activated-sludge process. Bacteria degrade the organic compounds in the water, creating a thick sludge that can be skimmed from the water. Tom's managers want to reduce the volume of sludge, which is expensive to dispose, by drying it, i.e., removing more water. Tom is going to try a new process that combines filtration with electricity, which laboratory experiments have shown to be an advantage.

To perform the test, Tom must set up a pilot plant—a small unit, but one big enough to provide engineering numbers through which a full-scale unit can be designed. Pilot plants are critical to new process development; they save the expense of a full-blown unit but provide more details than laboratory test-tube experiments. The equipment cost of Tom's pilot plant will be about $100,000, but a full-size plant will eventually cost about $5 million.

Tom's specific problem on this day is to figure out how to measure the electrochemical values of the sludge mixture. This is tricky because these values are usually measured in a liquid, not in a thick sludge. Tom consults with an instrumentation engineer (who happens to have an electrical engineering degree) and learns of the various types of instruments that can be used. He and his boss also review the *flowsheet*—a one-page drawing of the essential elements of the process. They determine that the best place to put the instrument will be near the entrance to the vessel where the sludge stream comes in to be filtered and dried. By performing some calculations on the fluid flow near that section of the vessel, they conclude that the sludge will be vigorously mixed, which in turn will mean that the electrochemical instrument will take good readings.

After several months' more work, the test results begin to come in from the pilot plant, and they're good. The addition of an electric field helps to remove 25 percent more water from the sludge. Tom now does a cost analysis, obtaining the price for disposing the sludge on a per-pound basis. With more water removed, the weight of the sludge (relative to the purified water) is lower, reducing disposal costs. On the minus side, however, the cost of electric power and the price of the treatment equipment must be totaled. Overall, the process will reduce water treatment costs by 15 percent, which is sufficient justification for building the plant. Tom writes up a report and now gets ready to deliver it in a presentation to upper management.

Education

Chemical engineering is one of the more intensive engineering programs to study. In addition to learning the basics of engineering, students must also carry six courses in chemistry—almost as much as a full-time chemistry student must attend. These chemistry courses include general chemistry, which nearly all engineering students take; organic chemistry; and physical chemistry.

Specific chemical engineering courses include:

- mass and energy balances
- thermodynamics
- process design
- transport phenomena—the concept of fluids moving through an area and the changes that occur
- chemical engineering economics

In addition a number of technical electives are usually required. Students seeking to work in biotechnology or pharmaceuticals, for example, might consider taking biology courses. Others may take electives in computer science, materials science, geology, polymer chemistry, or energy.

MINING, METALLURGICAL, AND PETROLEUM ENGINEERING

No smokestacks are smokier, no earthmovers leave bigger scars, than those used by mining and metallurgical engineers. The mines, steel mills, and ore refining plants that these engineers design and manage have a major impact on the environment, and thus environmental sensitivities are ever more important to these professions. By the same token, perhaps no other engineering disciplines are so vital to the manufacturing might of the United States.

It remains truer than ever that the fundamental activity of creating value from natural resources is the path to economic well-being for a country. The majority of American workers are employed in services: running banks, selling consumer goods, providing medical care. These activities represent a trade between a buyer and a seller, and no matter how many times they are repeated, the size of the American economic pie does not increase. However, extracting ore from the ground, turning it into semifinished sheets or bars, and then fabricating these into finished goods does generate wealth. (The other fundamental activity that generates wealth is farming and agriculture.)

Mining and metallurgy are linked because many firms dig the ore and minerals from the ground as well as refine them into ingots. In the case of steel, the intermediary step of refining ore into iron, and then processing that into steel, occurs. And metallurgy is one of a group of materials sciences, including ceramics, polymers, and fibers (textiles). These industries have a long tradition—even going back to before recorded history. In more recent years, technological change has altered how these engineering disciplines are taught. A student used to be able to choose mining and metallurgical engineering as a major. But in the middle of this century, other materials, especially plastics, became important. As a result new engineering disciplines such as polymer engineering and textile engineering came into being. Along the way ceramic engineering developed. This is the study of glasses; nonmetallic minerals, such as clay; and stone. Some schools have combined the entire group together in departments of materials science and engineering.

Thus, today there are a variety of engineering degrees to be obtained. The figures for graduating-class size from 1993 give an indication of how student interest is positioned; 1988 data show how it changed over those five years:

Table 2.   B.S.E. Class Size in Ceramics,
Materials, and Mining

	1988	1993
ceramic	368	286
materials & metallurgical	877	944
mining & geological	404	293
TOTAL	1,649	1,523

As this chart shows, ceramic and mining engineering have declined in recent years, while materials and metallurgical engineering has increased slightly.

Over the long term, federal data from the Bureau of Labor Statistics show that the overall materials field will grow by about 17 percent between 1988 and the year 2000—a couple of percentage points below the average for all professions, but still a positive sign.

History

The history of mining and metallurgical engineering is one of the factors that helps define human civilization; archaeologists date the transition from the Stone Age to the Bronze Age to the Iron Age starting about five thousand years ago.

Mining and metal refining developed gradually over the centuries, spurred by the discoveries of new metals, the drive for precious metal refining (gold and silver), and the needs of developing technologies. In the New World the availability of gold and silver in South America was one of the driving forces for Spanish conquest. In North America, although there were many searches for precious metals, little happened in this regard until the mid-1800s, when gold was discovered in California and the West. Coal mining, on the other hand, was a major industry throughout that century. Huge iron ore deposits in Minnesota (the Mesabi Range, which is still being mined) helped get the newly invented steel industry off and running in the late 1800s. Henry Bessemer, an English scientist, is generally credited with inventing the Bessemer process for producing low-carbon iron that could readily be converted to steel.

Production methods for other metals and minerals were being developed rapidly in the late 1800s, including techniques for producing aluminum, copper, zinc, and lead; and glass, gypsum plaster, and cement for concrete. A method for strengthening rubber made it more valuable as a material for tires; the young polymer industry also began to grow, spurred primarily by methods for making fibers from wood pulp.

These events helped establish the professional status of engineers specializing in metals and minerals. The Society of Mining Engineers was founded in 1871; the American Society of Metals (now known as ASM International) began in 1913; and the American Ceramic Society formed in 1898.

As the American West began to be explored in earnest in the latter half of the 1800s, many new sources of metal were found, including copper in Arizona and Idaho, lead and zinc in Colorado, and aluminum in Washington and Oregon. But these sources were limited in size and purity, and the search went around the world. By the mid-1900s, important sources of ore for American industry were in South America, Africa, Canada, and the Far East.

In the twentieth century the emergence of polymer technology, primarily plastics, began to change the mix of materials used by industry. The automotive and aircraft industries created a need for newer, higher-strength metals. The use of electricity to refine metals grew in importance. These trends led to the creation of the Society for the Advancement of Material and Process Engineering (SAMPE) in 1943. Finally, beginning in the 1950s but growing in intensity with each passing year, the needs of the microelectronics industry for ultrapure silicon, exotic new minerals such as gallium arsenide, or rare-earth minerals such as tellurium or yttrium, spurred new technology for minerals processing.

The Current Scene

After years of declining prospects, both mining and basic metalmaking have undergone a resurgence in the past ten years. There are a number of reasons for this renewal—most significantly, the development of new technology that allows the recovery of poor, low-grade ores economically. Metal refiners have made great strides in reducing the pollution generated during the refining process and in improving energy efficiency. Although imports are a problem for the industry, many engineers' careers are boosted by the opportunities for overseas assignments at the mining and refining operations of multinational firms.

Most ceramics are used in construction—cement and concrete, wallboard, and glass windows. New technology has affected sections of the industry from time to time, but imports are not as much of a problem. Rather, the ups and downs of the construction business affect industrial growth. For most of the 1980s, this growth was strong, but lately it has begun to settle down.

Both ceramics and metals are affected by the other major material—polymers. Materials engineeers and chemical engineers share this technological territory. Many of the major producers of polymer raw materials are also the producers of finished goods such as textiles or composite structures.

As polymers increase in strength and durability, they offer a lower-cost substitute for glass and metal. More and more bottles for foods and beverages are now being fabricated from plastic. The automotive industry is increasing the plastic content of cars and reducing the metal content. All-plastic airplanes are on the drawing board. The dramatic Stealth bomber that was unveiled in 1988 by Northrop Corporation features wings and a fuselage made from high-tech composites that combine glass fibers with plastic resins. The use of plastic makes the aircraft lighter and less visible to radar sensing.

The materials industries jumped in prominence a couple of years ago with the discovery of "warm" superconductors. These are a combination of copper and rare-earth minerals that can conduct electricity with no power loss. Previous superconductors needed to be kept at nearly absolute zero temperature—a freezing -460° F. It may be possible to develop superconductors that can carry current at room temperature, thus reducing the cost of conveying it.

Today the most rapid advances in materials technology are occurring in the aerospace industry and in microelectronics production. But the old-line metallurgy businesses also show new life. In the steel industry, for example, it is now common to build minimills that don't process raw iron, but use scrap steel instead. As the name implies, this practice dramatically reduces the size of a mill.

That minimills succeed in using scrap steel is a message to other materials producers; recycling and reuse are in. The growing shortage of disposal space for garbage and waste means that more materials recycling must be established. Aluminum, glass, copper, lead, and steel producers are already good recyclers. Now plastics and paper manufacturers are cranking up capacity and technology for recycling these materials.

Job Titles

The structure of job classifications varies widely from industry to industry in materials, mining, and metallurgy. Here are some of the more prominent job titles:

Plant superintendent. Mines and metal smelters are geared specifically toward producing the desired metal or mineral. The key technical person is the superintendent, who oversees production, labor, and costs. There is constant pressure to keep the mills running as various types of ores are brought in, or as the mine's topography changes in following a rich vein of ore.

Environmental manager. Because mining, metals, and materials production have a high potential for pollution, extra efforts are expended on environmental control. These include treatment of exhaust gases from smokestacks, water treatment of process water, and disposal of wastes or residues.

Development engineer. Microelectronic chips are produced on a "fab" (fabrication) line where platters of pure silicon are successively painted, etched, screened, or subjected to radiated light or molecular beams. The equipment to perform these functions is highly complex, and engineers are often called on both to run the machines and to help improve their performance. As opposed to a smoky factory floor, workers in fab lines must keep the air so clean that moonsuits (a head-to-toe covering) are worn, and the air is continually purified.

Specifications engineer. In construction companies, aerospace or automotive producers, and other materials-using industries, engineers are needed to check the desired specifications of a component that will be used in a larger assemblage. The appropriate material may be a corrosion-resistant stainless steel, a new high-strength plastic, or even a crack-resistant ceramic.

Melt Shop Manager

The heart of a steel mill is the melt shop—the location where a furnace liquefies metal in a searingly hot vessel. The chemical composition of the "melt," as it is called, is then checked and adjusted, and the molten metal is poured out to form ingots. The ingots are then flattened, smoothed, or cut into standard shapes.

But Tom is the proud engineer responsible for developing a revolutionary new method of making steel called thin-slab continuous casting. A thin slab is about an inch thick. It is now possible, and conventional, to cast slabs that are six or eight inches thick in a continuous caster. From there the slab is rolled to thinner dimensions in a section of the plant called a "hot strip." Finally it is cold-rolled to form sheet steel. Thousands of tons of coiled sheet steel are sold annually to auto and appliance makers to produce their products. If Tom's project succeeds, it will eliminate most of the hot strip, which would dramatically reduce the cost of producing sheet steel. The big problem, though, is to produce steel with few flaws or impurities because this high-quality steel is being used on the outer surfaces of products. No one would buy a new car with holes or pieces of slag on the fenders.

Tom is not alone in this effort; the entire company management is watching its progress daily. Today Tom's challenge is to improve the heat retention of the molten steel as it comes out of the furnace. If it cools too quickly, cracks form as it is being drawn into the thin slab. To research the problem, Tom has set a couple of dozen thermocouples (high-temperature thermometers) in the channel where the molten steel flows. As it goes by, he gets a set of readings of the temperature in each section of the channel.

Tom finds that the middle of the channel is hot enough, but that the sides are too cool, sometimes even causing solidified steel to stick to them. There are two choices readily available: adding heat to the sides of the channel, or insulating it better so that it doesn't cool as rapidly. Both have their pros and cons, but Tom senses that insulation may be the more dependable, and less expensive, route to go. He now begins researching the variety of insulating bricks that are available. He will get a sample of each material from their manufacturers and then run tests on all of them.

Education

Mining, metallurgical, ceramics, and polymer engineering cover a wide range of technologies that engineers must understand. Engineers must learn engineering basics first and then become familiar with the particulars of the chemistry of metals and minerals. Mining engineers must take structural and construction courses comparable to civil engineers in order to understand how to build mines and quarries. Polymer engineers study organic chemistry to understand the chemical principles of carbon, the primary component of plastics.

Not all engineering departments try, or want, to cram all the necessary study into a four-year undergraduate program. Five-year programs leading to a professional or master's-level degree are common. Two academic tracks have developed for many materials engineers: a bachelor's degree that can lead to a job in production and a master's or Ph.D. degree for design, research, and development. Courses for these programs include:

- organic chemistry
- soil science and geology
- metallurgy
- strength of materials
- polymer science
- thermodynamics
- mechanical design
- microelectronics fabrication

CHAPTER 4

MODERN ENGINEERING

This chapter is so titled not because the disciplines in other chapters are old-fashioned, but simply because all of the professions discussed in this chapter evolved in the twentieth century. The largest of the disciplines in this chapter, industrial/manufacturing engineering, originated in the growing complexity of production technology. A key milestone was the development of the assembly line at Ford Motor Company, where new techniques allowed a complete car to flow out of the factory every couple of minutes. For the time, the development was revolutionary.

A similar story could be told of other engineering disciplines. Aerospace/aeronautical engineering arose from the new technology of heavier-than-air flight. Petroleum engineering stepped into the spotlight as the needs of the internal combustion engine (especially in automobiles) ballooned in the first couple of decades of the century. The computer became a reality shortly after World War II and has since become a dominant force in our society. Finally the growing concentration of manufacturers, cities, and homes has created the need for environmental engineering—adapting technology to preventing pollution of the air, water, and land.

INDUSTRIAL/MANUFACTURING ENGINEERING

One of the most important words to manufacturers is productivity—the ratio between hours of work expended and the volume of products made or work performed. What is the best way to arrange the elements of a manufacturing process to achieve the highest productivity? That is the question industrial and manufacturing engineers address on a daily basis.

There is a strong overlap between two types of engineers involved with manufacturing productivity. The industrial engineer is a graduate of a four-year program with the same academic background as the other engineers listed in the previous chapter. Most manufacturing engineers, on the other hand, are graduates of programs in engineering technology. There are some industrial engineers called manufacturing engineers; conversely, it is possible to get a degree in industrial engineering technology. The big difference between the two is the type of undergraduate education engineers and engineering technologists receive. Engineering technology is more applied learning with courses targeted toward specific activities or industries. Industrial engineers receive a broader, more theoretically oriented education. Thus, while both can and do work for manufacturers, most technologists are restricted to that field of employment. Industrial engineers can find work at hospitals, construction firms, transportation services, and business consulting firms. This section will focus on all industrial engineers, as well as engineering technologists who focus on manufacturing.

Combined, these engineers and technologists number roughly 119,000, according to the Bureau of Labor Statistics. They project that the field will grow by 17 percent—slightly above the average for all professions—by the year 2005.

History

Industrial engineering traces its roots back to the new complexities of manufacturing on assembly lines, which were first organized for automobiles and agricultural harvesters and then extended to textiles, food processing, and the assembly of fabricated parts of all types. In the 1910s and 1920s, the field became formalized as "scientific management," in the belief that some fundamentally new ways of managing workers had been created that had a stronger mathematical base than the seat-of-the-pants management styles of the past. Frederick Taylor was the leading developer of this philosophy.

Manufacturing complexity continued to increase, but the Great Depression interfered with any major technological progress. Machine tool engineers and machinists in the Detroit area gathered in 1932 to see what could be done. They decided to organize what was then called the American Society of Tool Engineers, which evolved in 1960 into the American Society of Tool and Manufacturing Engineers. The name was again changed in 1970 to the Society of Manufacturing Engineers (SME).

During World War II, the need for enormous amounts of war materials sent a jolt through industry. Out of this heightened interest in improving manufacturing productivity, the Institute of Industrial Engineering was formed in 1948. Industrial growth and productivity boomed during the 1950s and 1960s, but hit some snags in the 1970s. By the end of that decade, the woeful state of manufacturing in the United States relative to other countries in the world became apparent as imports flooded the steel, automobile, consumer electronics, and other industries. Since then, American industry has been on a productivity and quality-improvement kick. This drive has opened up many new opportunities for industrial and manufacturing engineers. Between 1977 and 1987, the membership of SME nearly doubled, to about 80,000.

Industrial Specialties

The quality and productivity battles that American industry is waging against the rest of the world's manufacturers are still going strong. Nearly all manufacturers are looking closely at their production lines and workforces. A technological background is important for nearly all types of manufacturing managers.

The Institute of Industrial Engineers is organized into three societies (the Society for Health Systems; for Engineering and Management Systems; and the Aerospace and Defense Society), eight divisions, and nine interest groups:

Interest Groups
Computer and Information Systems
Consultants
Electronics Industry
Engineering Design
Government
Process Industries
Production and Inventory Control
Retail
Logistics, Transportation, and Distribution

Divisions
Energy, Environment, and Plant Engineering
Engineering Economy
Facilities Planning and Design
Financial Services
Operations Research
Quality Control and Reliability
Utilities
Ergonomics and Work Measurement

The latter organization, in a way, represents the classical scientific management that is the heart of industrial engineering. In order to maximize productivity, industrial engineers will stand by workers on an assembly line with a stopwatch and record the motions workers make as they complete tasks. This time and motion study is then translated into the procedures that all workers are to follow. Most recently American firms have adopted the model of European factories. One or several workers are responsible for all the assembly steps of a unit or component, rather than simply screwing one part onto another and passing the result down the line.

The Society of Manufacturing Engineers divides its membership into fifteen technical activity areas, based on the specific manufacturing technology involved. These include material forming, finishing, robotics, sensors, automated fastening, composites, and manufacturing management.

Job titles range from production manager to quality assurance engineer to control engineer. Because both industrial and manufacturing engineers are so intimately involved with workers and plant management, this background can be a sturdy stepping-stone into corporate management.

Industrial engineers study the normal load of fundamental engineering courses and then select from a variety of specialties such as labor relations, industrial psychology, computers, economics, and business management.

AEROSPACE/AERONAUTICAL ENGINEERING

To fly! That ancient dream became a reality at Kitty Hawk, North Carolina, in the dawning days of this century. Now, more than eighty years later, flying can be done in a bewildering array of balloons, satellites, rockets, airplanes, helicopters, jetliners, and gliders.

The technical people responsible for keeping all these devices aloft, and those who design and build new types, are aerospace/aeronautical engineers. (In this book the term aerospace will be used to cover both of these terms.) Aerospace engineers aren't the only ones who work on aircraft; mechanical, electrical, computer, and materials engineers are also well represented. However, aerospace engineers have the central responsibility of designing the shape and performance characteristics of the craft and of specifying a propulsion system and guidance controls.

There are about 66,000 aerospace engineers at work today, according to federal data; the number is projected to rise by 14 percent by the year 2005. That percentage is considerably less than the average for all professions.

History

The federal government has been a prime mover of aerospace technology throughout this century. After the inaugural flight of the Wright brothers, the government began producing planes for the army, which wanted to use them for battlefield surveillance. The United States Post Office also became involved, seeing airplanes as an obvious delivery vehicle for the mail. Around this time, William Goddard, a scientist, was working in near-total obscurity on rocket technology—a field that the Defense Department practically ignored until after World War II.

Enough of an airplane industry had developed by the early 1930s to warrant the establishment of the American Institute of Aeronautics and Astronautics, Inc. (AIAA), in 1932. Today the Institute has about 40,000 members.

Aircraft design and construction took a huge jump during World War II, as all sorts of fighter aircraft, bombers, and cargo planes were developed. After the war some of these designs were adopted to civilian uses, such as the venerable Douglas DC-3. The development of the turbojet created a new class of aircraft and opened a new era of international travel. The term "jet set" was coined in the early 1960s as a reference to the possibilities of world travel.

In the 1960s the American space program hit its stride with the objective of landing a man on the moon before the end of the decade. This happened in 1969, and it was followed by a handful of later flights. Spacelab was built (later to fall to earth), and then the major portion of funding from the National Aeronautics and Space Administration (NASA) was shifted to the Space Shuttle, which is in use today.

Also after World War II, the rocket became a critical component in the nuclear defense arsenal of the United States. Rockets were also the main method of lofting satellites into space, especially communications satellites that permitted live, around-the-world transmissions.

The end of the Cold War between the United States and the former Soviet Union has had a dramatic impact on the aerospace engineering profession. The United States is now "building down" (a euphemism for reducing) the size of its military forces, cutting back on the production of aircraft, and stretching out the development cycle for new-generation aircraft. Although not all aerospace engineers have been employed directly in making aircraft for the military, the many that were have cut off many opportunities for newly graduated engineers. The main defense contractors are consolidating and reducing their staffs. While the greatest part of this shrinkage has already occurred (and will surely have ended by the time today's high school students graduate from college), the profession will look different in coming years. More work will be obtained in civilian aircraft and in satellite technologies for communications.

Specializations

Aerospace engineers have learned a tremendous amount about the performance of materials and structures under extreme stress. This knowledge is required for developing new aircraft, and it can also be adapted to other fields such as land-based transportation, power generators, electronic controls, and related areas. Thus, though most aerospace engineers work for large government contractors, their expertise can be applied in many other areas.

AIAA identifies seven areas of concentration among its members. They are as follows:

1. Propulsion. The means by which aircraft move onward and upward.
2. Fluid Mechanics. The fluid in this case is air, flowing into jet engines or over wings.
3. Thermodynamics. Handling extremes of heat and cold in an aircraft—or even the cold vacuum of space.
4. Structures. Aircraft must be able to withstand the intense stresses of takeoff, flight, and landing.
5. Celestial Mechanics. How does one predict that a satellite launched this year will pass close by Saturn three or five years from now? Celestial mechanics provides the answers.
6. Acoustics. Acoustics—the study of sound—is important to prevent vibration and noise pollution from aircraft.
7. Guidance and Control. Another obvious element of aircraft design. Some of today's aircraft are so complex that a pilot cannot fly the plane alone; an onboard computer helps by making rapid adjustments.

Training for aerospace engineering includes the standard courses all engineering students take, plus a list of specialized courses including aerodynamic design, advanced mathematics, fluid mechanics, electronics, propulsion systems, trajectory dynamics, and structural analysis. Nearly one out of three aerospace engineers goes on to get a master's degree.

MATERIALS ENGINEERING

Petroleum engineering is a one-industry profession involving oil and gas production companies. For this reason, when the oil business is good, job opportunities for petroleum engineers are *very* good; and when business is bad, opportunities are *very* bad. In the late 1970s and early 1980s, when the price of crude oil zoomed to $38 per barrel, petroleum engineering was a magnet for students. Since then, however, the price has dropped to around $18 per barrel. Oil companies have been contracting and reducing the amount of exploration they perform. Still, what goes down must come up when it comes to oil. Most of the major producers are hiring again, and there is hardly anyone who foresees a further contraction of the industry.

History

The first well drilled to produce crude oil was in Titusville, Pennsylvania, in 1859. It was all of seventy feet deep. By comparison, wells are being drilled today through a half-mile of water (in the Gulf of Mexico), digging into the ground for over another five miles. Getting from the earliest well technology to the latest has been the work of the petroleum engineering profession. By and large it divides into two functions: exploring for oil and then producing it. Production can be simply drilling a hole and attaching a pump (or a throttling valve if the pressure of the field is so high that oil would gush forth); alternatively, petroleum engineers are learning how to use detergent chemicals and solvents to "scrub" oil out of what had been considered an exhausted well.

Today there are about 14,000 petroleum engineers, and the federal projection shows no growth over the coming decade.

Like the mining industry, much of today's oil production occurs away from the continental United States, which has been surveyed and drilled extensively over the past century. The Prudhoe Bay oilfield on the North Slope of Alaska was the last major oil find in the United States. Production of oil has dropped from just over ten million barrels a year in the early 1980s to fewer than eight million today. Further declines are expected.

Exploration and Production

The major American oil firms remain dominant forces in the world market for exploration and production. This is good news for engineering students, who have the opportunity to work abroad through one of these firms.

Exploring for, and producing, oil calls on a host of technologies. Exploration engineers will use seismographic techniques, satellite photographs, and underground rock samples to locate an oil pocket. A major development in this step has been the use of the latest supercomputers to process huge amounts of data on underground formations in order to locate reservoirs more easily.

Once a site has been selected, a drilling crew sets up a rig. The petroleum engineer monitors the types of rocks and soil the drill cuts through and can further test the site by running a "string" of electronic instruments down the well to analyze geological conditions. There is an interesting combination of outdoors work—sometimes in the most remote regions of the world—and high-technology tasks in computer-equipped offices.

Something like six out of seven wells is dry—not enough oil is found to make it worthwhile to extract it. For the lucky one well, though, the next step is to bring in a pipeline and associated pumping equipment to convey the oil back to a central refinery.

Education

Education for petroleum engineering combines several aspects of civil engineering, chemical engineering, and geology. Students take courses in geology, hydrogeology, and chemistry to understand the dynamics of underground wells. There are more courses on reservoir engineering—the practices of maintaining a field's capacity once extraction begins.

ENVIRONMENTAL ENGINEERING

In the 1980s environmental engineering was one of the hottest careers going. Environmental protection didn't just "happen" in that decade; it was built on concerns and technologies that had been maturing for decades. But the momentum to stop damaging the environment and to repair previous injuries took on new intensity at that time, and new regulations spurred the growth of hundreds of companies. Since then, environmentalism has slowed considerably, but it remains at a much higher level than ten years ago.

Most engineers who work in the environmental field start by obtaining a civil engineering degree. This is fine, but many observers of the environmental field believe that a different curriculum should be taught. Specifically, they believe civil engineering should be combined with a knowledge of biology—microbes; ecological systems, such as forests or estuaries; and biochemistry. There are a limited number of schools where environmental engineering is taught under that specific title. More often, one takes an environmental concentration in civil, chemical, or mechanical engineering.

Estimates of the current size of the environmental engineering field are hard to come by because there is no one central organization. The Academy of Environmental Engineers, with about 2,500 members, represents the certified, experienced environmental professionals. Several thousand civil engineers within the American Society of Civil Engineers specialize in environmental issues; a similar case exists within the American Institute of Chemical Engineers.

Adding up all these numbers gives a total of around 20,000-30,000 environmental professionals. Only about 150 students obtain a B.S. degree in environmental engineering each year (another 350 or so earn master's degrees). Therefore, if the field is going to grow as rapidly as most people think, the new engineers will come from other programs.

In the mid-1980s the growth rate for some environmental engineering specialties ran at 20 percent per year. That kind of growth is unsustainable. Many manufacturers, who often bear the brunt of new regulations, had to rapidly step up their cleanup programs. But once those programs were in place, the need for new engineering talent waned. Current growth rates are a more sustainable 5–10 percent. The new deregulating philosophy that is taking shape in Washington may slow this down even further.

Environmental concern is nothing new. The Air Pollution Control Federation, whose 10,000 members are concerned with air pollution and solid wastes, was founded in 1907. But the growing crisis in the environment, with hazardous wastes, accidental exposures of toxic chemicals, railway accidents, and polluted rivers and streams, has cast environmental engineering in a new light.

Specifically it is now more appropriate to think about, and deal with, environmental pollution as a "multimedia" problem. This means that what might start as an air pollutant (dust from a smelter, say), can become a liquid slurry waste (in a stackgas filtration unit), and then, after drying, wind up as a solid waste. It is no longer meaningful to talk about air pollution separate from water or land pollution.

Job Opportunities

Today's environmental engineers work for local, state, and federal government and for engineering firms that perform consulting work for government agencies. Environmental engineers also work in industries, such as organic chemicals or metals extraction, that have the potential to create significant amounts of pollution. A study by Rutgers University in 1994 found that half of the environmental professionals working in industry were in either chemicals, petroleum and coal, or primary metals. Although some engineers in industry are able to specialize in the control of air pollution, water pollution, or solid waste, at many companies these responsibilities are merged into an environmental services department.

Job titles range widely, as does the type of employer. Moreover, the field is evolving rapidly, generating new types of jobs. Some of the more common ones in existence today are:

- *Site Manager* at EPA-sponsored hazardous-waste cleanup projects.
- *Compliance Officer* for the environmental-control systems at a manufacturing plant.
- *Design Engineer* at an engineering construction firm, responsible for developing environmental systems for private or public works.
- *Enforcement Official* for a state environmental agency or for the United States Environmental Protection Agency.

The career path for many environmental engineers begins at state or federal environmental agencies, which provide training in technology and regulatory issues. With this background many engineers then move on to higher-paying jobs in private industry.

COMPUTER ENGINEERING

The booming computer industry has two basic types of technical experts: engineers for computer hardware, and computer scientists and programmers for computer software. Midway between the two is the computer engineer who must be aware of trends at both ends of the scale.

Computer engineering is a relatively new field, having developed as an academic program over the past twenty years or so. Many students studying electrical engineering and specializing in microelectronics call themselves computer engineers, but the electrical engineering curriculum tends to emphasize hardware over software. Conversely it is possible to study computer science in college and graduate with an understanding of both hardware and software by taking microelectronics courses. Because of this overlap, many schools have begun offering degrees in computer engineering. The computer engineering curriculum offers more software courses than the normal electrical engineering major and more hardware courses than the average computer science major. Computer engineering is targeted more specifically at the computer industry.

The field has also been affected greatly by new technology. Only a dozen years ago, the number of computer configurations was fairly stable. There were large, fast mainframe computers for running many different operations rapidly, over large sets of data, for an entire corporation. There were minicomputers, which were smaller, slower, but less expensive and suitable for fulfilling the computer needs of a department. And the microcomputer, better known as the personal computer (PC), was being offered for the single user.

Since then, the supercomputer has evolved as a very fast number-crunching machine for scientific applications. The PC has become a major force in the computer world; there are something like eighty million of them currently in use in the United States. Meanwhile specialized computers have appeared for artificial-intelligence programming, enhanced graphics processing, and high-speed factory automation equipment. And the general structure of these computers ranges from single-processor (one central "brain" that performs computations) to multiprocessor designs. Some of the latter are set up for a form of computing called "parallel processing," which allows the computer to execute programs more rapidly.

Job Opportunities

All these styles, or "architectures," for computers have created a boom for people trained in computer design—computer engineers. The computer field is huge and dynamic. Jobs—even companies—are not stable. But it remains one of the most entrepreneurial fields of technology, one where careers can move upward rapidly. But it is also possible to have one's employer go out of business overnight.

Most computer engineers work at firms that design and build computers. Some also find opportunities in the much larger realm of computer users. Banks, insurance companies, universities, and research groups all have a need for engineers who thoroughly understand computer hardware, yet are also aware of how computers are being put to use. Computer engineers in the latter position help their employers set up suitable computer departments and help specify the networks, communication equipment, and printers for outputting data.

Job demand is very strong for the computer engineer. Students have been flocking to computer engineering programs. There were a thousand or so B.S. graduates in the late 1970s, and there are almost 5,000 graduates today. The business has a bright future.

CHAPTER 5

ENGINEERING SPECIALTIES

The many engineering disciplines listed so far by no means exhaust the various fields of study available. Whatever your interests, there is probably an engineering department devoted to that study, or to something that comes awfully close. Many schools also have an independent-study option that would allow you to exercise your curiosities.

The list of possible fields of study is greatly expanded at the master's degree level. So much has to be crammed into an undergraduate education; it is hard to develop both a broad, general knowledge of engineering and sufficient understanding of some specialized fields. You may think that going to graduate school would require too much time or money. Only about a quarter of all undergraduate engineering students go to graduate school. However, many of those students are earning their master's degree while already starting their working career. They do so by attending graduate school at night; this gives students an income while they are still getting trained. Many companies also cover the cost of tuition for graduate school.

The following are some of the better-known engineering specialties:

- biomedical/bioengineering
- agricultural engineering
- nuclear engineering

- marine/ocean engineering and naval architecture
- safety and fire protection engineering
- optical engineering
- automotive engineering
- textile engineering
- energy engineering
- heating, ventilating, and air-conditioning engineering
- systems engineering/operations research
- engineering history/technical writing

BIOMEDICAL/BIOENGINEERING

Biomedical engineering or bioengineering is developing into two types of specialties: the engineering design of body parts, especially human ones; and the application of the biotechnology revolution.

During the 1950s and 1960s, the growing body of knowledge about how engineering systems worked—structures, fluid flow, chemical reactions, electronics—led to a belief that the human body could be "engineered" in much the same way that a bridge is built or a pump is designed. Although the initial optimism has been reduced by the daunting challenges of building prosthetic devices, steady progress has been made. Here is a partial list of the replacement parts that are now in use or near commercialization:

bones and joints
cartilages
skin
heart
eye corneas
lung
kidney
teeth

blood vessels and blood
heart valves
hair
ears (hearing aids)
muscle controls (nerves)

A quiet revolution has occurred in the past fifteen years or so, as medical doctors and engineers collaborated on this growing list of prosthetics. Even more advances have been made in the area of analytical or diagnostic instruments. One example is computerized axial tomography (CAT) scanning, which can probe the human body to obtain visual diagrams of internal organs. Much of this type of work depends on top-notch electrical engineering.

Starting around 1980, genetic engineering arrived on the scene, creating a boom in pharmaceutical research that now goes under the general title of biotechnology. Engineers are rarely involved in actual genetic work. However, there are many applications involving advanced genetic manipulations in combination with manufacturing or research. These include:

- commercial-scale production of pharmaceutical agents
- industrial chemical production via microbes
- agricultural production with genetically altered plants
- production of sensors and diagnostic devices
- purification of biological compounds

There are a number of academic departments across the country where engineers can study these subjects. Some of them are available on an undergraduate basis; many of them are options or master's-level programs associated with chemical engineering departments. About 650 engineers received a B.S. in biomedical/bioengineering in 1988; there were also about 250 M.S.E. graduates. Most of these graduates head toward the pharmaceutical industry for jobs; many also go on to obtain a Ph.D.

AGRICULTURAL ENGINEERING

Agricultural engineering applies the lessons of the factory to the farm. The mechanization of agricultural work has been a constant process for more than a century in the United States; agricultural production continues to go up, but the number of actual farmers continues to decline.

The roots of agricultural engineering as a distinct profession go back to the beginning of the century when a group of mostly mechanical engineers organized the American Society of Agricultural Engineers (ASAE). That organization now has almost 10,000 members.

Like the founders of the profession, most agricultural engineers today are involved in developing machines and vehicles for farming. Soil conservation, forestry, and food production are also key areas. ASAE defines nine technical divisions among its members:

1. *Aquacultural Engineering,* the study of increased production of sealife such as fish farms and hatcheries.
2. *Bioengineering,* the application of biological science to plant and animal production.
3. *Electrical and Electronic Systems,* which ranges from electronic control of processing systems to instruments for measuring irrigation, feeding, and harvesting.
4. *Food and Process Engineering,* the application of engineering principles to the processing, handling, packaging, and storage of foodstuffs.
5. *Forest Engineering,* the enhancement of tree production and harvesting.
6. *Knowledge Systems,* the use of advanced computer programs such as artificial intelligence to aid in production and processing.
7. *Power and Machinery,* the traditional element of agricultural engineering focused on vehicle design and power supplies.

8. *Soil and Water,* which includes irrigation, drainage, fertilization, and water-resource management for agriculture.
9. *Structures and Environment,* which ranges from barn construction to the development of advanced feedlots for livestock or storage systems for crops.

As these examples show, agricultural engineering cuts across a broad range of engineering topics. The overall field has had moderately good job prospects for the past several years, notwithstanding the continuing grim news among farm owners. The number of homesteads and family farms has dropped precipitously over the past ten years and is predicted to continue to fall during the next decade. However, the amount of land for production—and jobs for engineers—has held fairly steady.

About 400 students received B.S.E. degrees in agricultural engineering in each of the past few years; that is down almost a third from the peak enrollments of the early 1980s. An additional 150 students earn M.S.E. degrees.

NUCLEAR ENGINEERING

Many commentators have spoken in recent years of the "extinction" of the nuclear industry in the United States following the accident at the Three Mile Island nuclear power plant in 1979. That accident caused the cancellation of dozens of construction projects; no new nuclear power plant has been ordered for construction in the United States since.

But the picture is much more complex than the ups and downs of the nuclear business among utility companies. First, the United States has a considerable involvement in plant construction abroad. Second, the U.S. Department of Defense continues to order new, smaller plants for use in nuclear-powered submarines and ships. Third, nuclear engineering also entails the design and construction of medical equipment for diagnosis and treatment of disease and for analytical instruments used in industry. Fourth, there are already about 110 nuclear power plants in the country; the operation, maintenance, and repair of them occupies the efforts of thousands of workers. Fifth, the Defense Department has sponsored an elaborate system of production facilities to purify radioactive materials for the production of nuclear weapons (including the concentrated fuel pellets that are the source of power for commercial nuclear plants). This system, parts of which can be traced back to the days of the Manhattan Project during World War II, is in desperate need of repair and refurbishing. The U.S. Department of Energy, which runs the system at the behest of the Defense Department, estimates that this renovation will cost more than $100 billion over the next decade or so. That's going to be a lot of work for nuclear and other engineers.

The foregoing represents fairly definite tasks that require nuclear engineering talents. There are also some possibilities for future growth. One possibility is that new nuclear plants will be ordered. Few people have been counting on this in recent years, but the sudden emergence of the greenhouse effect as a world-threatening pollution problem has changed some opinions. The greenhouse effect refers to the warming of the earth caused by increased levels of carbon dioxide (the product of any fuel's combustion) and other industrial byproducts. Nuclear plants, of course, don't burn fuel and therefore don't produce carbon dioxide. Also, the commercial utility network in the United States is rapidly running out of capacity, and a new round of plant construction will be necessary during the 1990s and beyond. These new plants don't have to be nuclear-based, but some of them could be.

Job Opportunities

Notwithstanding the strong career potential, fewer and fewer students are studying nuclear engineering. B.S. graduating class sizes peaked at around five hundred in the late 1970s, and have been slipping ever since. If nuclear power comes back into favor, there will be almost immediate demand for new engineers. The latest effort by the engineering design companies that build power plants are a variety of "inherently safe" designs that run at lower pressures or use design innovations developed in recent years to improve the safety of the plants.

Nuclear engineering is dominated by the power industry, but that isn't the exclusive employment opportunity. Nuclear medicine is a broad field, involving therapy for diseases and analytical techniques for detecting illness. Instruments using nuclear energy in some form are widely used in laboratories and in industry for maintenance or analysis of materials. These applications, though employing relatively few engineers, represent a broad range of opportunity.

Nuclear engineering students study the engineering basics, then additional courses in physics, power plant design, materials, and electronics. The 292 B.S.E. graduates in 1994 were complemented by 241 M.S.E. graduates. These graduates will find jobs available at the electric utilities, the military (as civilian engineers or as commissioned officers), at engineering/construction firms, environmental firms, and instrumentation manufacturers.

MARINE/OCEAN ENGINEERING AND NAVAL ARCHITECTURE

How much of American industry is on, in, or near the sea? Ships, of course, including the hundreds of craft that the United States Navy operates, are a major element. But there are also the offshore oil-drilling business, aquaculture, port design, undersea pipelines and telecommunications, dams, locks, and canals. Moreover, it is widely expected that the use of the world's oceans and waterways will increase in the future, if only because the technology to live or work in the ocean has only recently become available.

One example of this is the mining of naturally concentrated metallic nodules deep in ocean rifts. About a decade ago experimental submersible vessels discovered that there were volcanic vents that reached deep into the earth's core at the bottom of the Pacific Ocean. One day the mining of these nodules may be an important source of metals.

Another not-so-farfetched application concerns a technology known as OTEC, for ocean thermal energy conversion. Researchers have known for decades that the difference in temperature between the water at the ocean's surface and the water several hundred feet below could be used to drive energy-generating turbines. (It can be used simultaneously to produce salt-free water.) OTEC was experimented with during the days of high-priced energy in the 1970s. It was found to be possible but too expensive at that time. If energy prices rise in the future, OTEC may once again look attractive.

It is ironic to speak of such futuristic technologies while discussing naval architecture, since this profession is centuries old. People have built ships since before the beginning of recorded history. Today, in the United States, ship design and construction is relatively quiet, except for the military. The United States Navy has some five hundred vessels, but there is talk today of decommissioning many of them and replacing only a few. The market for recreational ships, such as motorboats and sailing yachts, is also relatively small, but steadier in employment prospects.

These three engineering disciplines—marine and ocean engineering, and naval architecture—have subtle differences in their outlook:

Marine engineering is generally concerned with mechanical systems on board ships, such as the propulsion system, controls, and heating and cooling. Some of this technology also applies to stationary equipment at or in the ocean.

Naval architecture is more concerned with the design and construction of hulls. Knowledge of the fluid dynamics of a vessel coursing through water is essential. Obviously there is a strong overlap between marine engineering and naval architecture. The distinctions between them are similar to those between mechanical engineers who design jet turbines and aerospace engineers who design aircraft powered by the turbines.

Ocean engineering has been likened to civil engineering with wet feet. Ocean engineers are concerned with structures—such as ports, drilling platforms, or pipelines—next to or in oceans. Basic construction technology must be studied, along with the special effects of tides, saltwater, and sealife on ocean structures.

Job Opportunities and Education

Shipbuilding and ocean exploration are international businesses, with firms on many different continents and opportunities to travel the world. These businesses are also highly competitive, and the United States maritime industry has not fared well outside of military contracting. Most shipbuilding for freight transportation, for example, is carried out in the Far East. Nevertheless, as the volume of shipping increases due to greater international trade, the industry could revive in coming years.

The training for marine engineering and naval architecture covers engineering fundamentals, as well as fluid dynamics, energy and propulsion systems, and control. Marine engineers have an edge over naval architects in the job market because maritime engineers receive training in propulsion that can be transferred to utility plants or other land-based power systems. Ocean engineering is often closely associated with civil engineering programs, and the studies are similar, involving materials and structures and construction technology.

Most schools for marine/ocean engineering and naval architecture, naturally enough, are in states on the nation's shores—the West and East Coasts and the Gulf of Mexico. The United States Navy provides employment opportunities and training for its engineering officers. It also hires a large number of engineers for civilian employment at its major ports, including Newport News, Virginia, and San Diego, California.

SAFETY AND FIRE PROTECTION ENGINEERING

Risk and technology often go hand in hand. Many problems that engineering addresses involve dangers that must be considered before a design is completed or a building is built. A good example can be found in the chemical industry, where chemical businesses routinely work with hazardous intermediate chemicals in order to produce a final product that might be as safe as water.

Good engineering and safe engineering are the same thing. Nevertheless, many engineered processes or products are improved by having a safety specialist involved in the planning. Many safety engineers work for, or are closely associated with, the insurance business. Before providing an insurance policy, an insurance firm will often have safety specialists review a business's factories or products, making recommendations to the client to improve the safety of the process or product. In the short term, this analysis can result in lower insurance premiums; in the long term, it can result in fewer accidents, a safer workplace, and a more profitable business.

There are relatively few engineering departments in the country that offer specific courses in safety engineering. Usually they are part of industrial engineering or engineering management programs. There are also a variety of programs in safety from such nonengineering departments as industrial relations. After graduation there are numerous programs sponsored by insurance companies, research foundations, and professional organizations such as the American Society of Safety Engineers (ASSE). Safety engineering courses focus on such areas as accident prevention, the design of safe manufacturing equipment, and industrial codes and standards.

Interest in safety engineering has increased greatly over the past fifteen years or so. The United States Occupational Safety and Health Administration (OSHA) put federal enforcement behind national workplace safety rules. (OSHA is also a good place to gain work experience.) More recently the explosive growth in insurance costs and insurance awards has made safety a conscious issue with many industrial managers. ASSE has grown from its traditional size of around 2,000 members to 20,000 members.

Safety engineering work requires a close understanding of the general rules of safe function or operation and familiarity with thousands of rules and technical standards for proper procedures. Research into such topics as inherently safe processes and human-machine interfaces further boosts the professionalism of the field.

A field closely related to safety engineering is fire protection engineering. These engineers, following study in any of a number of B.S. engineering programs, can study fire protection engineering at several schools that offer master's degree programs. Employment opportunities range from chemical and petrochemical plants, to insurance companies, to suppliers of fire safety equipment for buildings, homes, and factories.

OPTICAL ENGINEERING

Lasers. Fiber optics. Space telescopes. Microwave communications. Microelectronics fabrication. The list of applications of optics in the modern world is long and growing. As yet, there are only a few B.S. programs and a few more M.S. or Ph.D. programs at schools for optical engineering. But there are quite a few engineers and scientists who specialize in the field. There are about 7,000 members in the International Society for Optical Engineering and about 10,000 in the Optical Society of America.

Optical engineering involves understanding and working with the properties of light. New materials, such as ruby lasers and polymeric optical fibers, make it possible to generate new forms of light and to carry that light along wires.

The optical industry is closely allied with electronics and electrical engineering, since so many of the applications are in microelectronics and instrumentation. Students who want to pursue a career in optical engineering are advised to take extra courses in mathematics and physics in high school and during their undergraduate years. The best job opportunities develop after obtaining a master's or higher degree.

Much optical work is done by scientists, primarily physicists, who investigate new ways to produce or use light. Optical engineers then transform this research into commercial products. The job outlook is excellent, since many of the latest advances in microelectronics depend on optical principles.

OTHER ENGINEERING SPECIALTIES

The fields of study mentioned above still do not complete the list of engineering professions. The following specialties tend to be very small or have only limited presence on college campuses. In the latter case students usually major in one of the larger engineering disciplines in college and then specialize through professional affiliation with an engineering society. Often the engineering society will offer technical seminars or continuing education courses that provide the necessary training for the specialty. There are many specialties that meet these criteria.

Automotive Engineering

This field is not taught at any accredited college, yet the Society of Automotive Engineers has nearly 50,000 members—a very large group in engineering circles. The existence of this professional organization is a testament to the size and importance of the automotive and transportation industries in the United States. The automobile dominates the field in terms of professional interest. These engineers are concerned with engines, structural components, fuels, lubricants, suspensions, and related topics. Because these elements are common to other vehicles, automotive engineers also work in the aerospace industry or for manufacturers of heavy-construction equipment, farm vehicles, mass-transit systems, and trucks.

Textile Engineering

With several programs, mostly in schools in the South, textile engineers are trained to manage the design and operation of the weaving and cutting equipment that textile manufacturers use. Textiles, as a material, involve much more than clothing. One can also find textile engineers at high-tech aerospace firms, where they are concerned with handling the glass or carbon fibers that are used with plastic resins to make composite materials such as fiberglass.

Energy Engineering

The energy industry is huge, encompassing oil producers, utilities, power systems, solar energy production, batteries for electric storage, and many other forms of energy. Mechanical, electrical, and industrial engineers are well represented. Most energy engineers, on the other hand, are concerned with the use and conservation of energy. They might help develop plans to insulate a building, the better to conserve the energy needed to heat and cool it; or they might assist in a factory's plan to build its own power station, rather than buying power from a utility. The Association of Energy Engineers, with about 6,000 members, is the leading professional organization.

Heating, Ventilating, Air-Conditioning, and Refrigeration Engineering

This long name, usually abbreviated as HVACR, refers to the climate control systems of homes, buildings, and large vehicles. HVACR engineers, like energy engineers, are concerned with the use and conservation of energy. HVACR engineers also help create the conditions that make people comfortable and safe—an environment with the right temperature, humidity, cleanliness, and lighting. These concerns usually fall under the technical term *psychometrics*. Most HVACR engineers work for construction companies, building management firms, or equipment suppliers. The American Society of Heating, Refrigeration, and Air-Conditioning Engineers, Inc. has nearly 60,000 members.

Systems Engineeering/Operations Research

These two programs exist at many colleges, but there are relatively few professionals who identify themselves as systems engineers or operations researchers. Both programs tend to be more emphasized at M.S. or Ph.D. levels.

The field came about as scientists and engineers realized over the past century that a different skill is needed when a large, complex organization or structure is involved. An airline is a good example. There are mechanical engineers for the aircraft themselves and transportation engineers for airports, but who pulls together all the necessary elements—people, planes, baggage, fuel and spare parts, food and services? Someone must handle the large volume of customers and get the maximum profit. This is one example of the problems that systems engineering and operations research address.

Other topics include space or satellite programs, the design of new factories or production lines, or military equipment programs. The skills employed include computer programming and mathematical theory, combined with experimentation and observation. There are about 7,000 members of the Operations Research Society of America.

Engineering History/Technical Writing

The history of science and technology is intricate and fascinating, and it has only recently become popular on college campuses. Most teachers of the history of technology have undergraduate degrees in history or another of the liberal arts, but the advantages of having engineering training are obvious. Aside from academic positions at colleges and universities, historians of technology are employed by major corporations to write company histories and by journals and periodicals. The Society for the History of Technology has about 3,000 members.

Technical writing is similar to engineering history in that the main activity is writing, and also in that most professionals do not have an engineering degree (however, it is a benefit to have one). Technical writing covers a broad range of topics, including journalism, advertising copywriting, documentation for user's manuals, and contract preparation. A large segment of the technical writing field today is devoted to providing the documentation that goes along with computer programs. Most programs are not self-explanatory; they require a manual that guides the customer through the program steps. The Society for Technical Communications, Inc. has about 10,000 members.

CHAPTER 6

ENGINEERING TECHNOLOGY

Engineering technology (E.T.), sometimes called industrial technology, is becoming much more prominent today. As increased technology is employed in more parts of the American economy, more people with technical training are needed. According to the federal government's Bureau of Labor Statistics, job-growth projections to the year 2005 show that the large numbers of engineering technologists—more than 1.2 million—will increase at about the same rate as the workforce overall. Some 229,000 additional technologists will be working by then.

Most of the major engineering disciplines—mechanical, electrical, computer, civil, chemical, aerospace, and manufacturing—have a counterpart degree in technology. There are programs and jobs for mechanical engineering technologists, electrical engineering technologists, and so forth. Most of the details of the various engineering disciplines in the preceding chapters also hold for engineering technologists. Therefore, only the general situation for E.T. will be reviewed, with some examples of specific types.

THE PROS AND CONS OF
ENGINEERING TECHNOLOGY

The formal definition of engineering technology, as espoused by the American Society for Engineering Education, is as follows:

123

Engineering technology is that part of the technological field which requires the application of scientific and engineering knowledge and methods combined with technical skills in support of engineering activities; it lies in the area between the craftsman and the engineer in the part closest to the engineer.

Engineering technology makes sense for some people:

- Those who don't have the time or money to attend college full-time for four years.
- Those who enjoy science, math, and technology but don't feel confident enough of their aptitude in these subjects to study engineering.
- Those who want to work on the maintenance or operation of technical equipment, but are sure that they don't want to work on the design of such equipment or on research for new types of equipment.

In general, E.T. degrees are available at associate (two-year) and bachelor (four-year) levels. Many people define two-year graduates as technicians and four-year graduates as technologists. In reality the job titles may be intermingled. In fact the technologist may wind up being called an engineer.

There are a few programs that offer master's degrees, but only a very small number of students obtain them. Many two-year E.T. programs are offered at community or junior colleges. Many programs are also offered with full or partial sponsorship of local employers who can provide co-op programs that combine work with study.

In general the curriculum for engineering technology has basic math and science courses and then a number of courses tailored to specific industrial applications such as electrical systems, electronic components, machine tools, instrumentation, construction, or manufacturing methods. With this training, graduates can move directly into jobs as technicians, production personnel, or service specialists in a wide range of industries. Many employers like the idea of having an employee who is ready to work once hired; often, graduates of four-year engineering programs need a certain amount of on-the-job training before they begin to be productive.

Lacking the more comprehensive courses in higher math and science, most E.T. graduates are not equipped to work as design or development engineers. With experience and additional training, it is possible for technologists to get design work, if they so desire. And, the option of taking additional schooling to get the full engineering degree is always available. No doors are slammed shut because of prior experience as a technologist.

In one area, technologists and engineers are often competing for the same job: manufacturing supervision. Many E.T. graduates work as manufacturing specialists, just as graduates of programs in industrial or mechanical engineering do. The Society of Manufacturing Engineers, many of whose members have an E.T. degree, has nearly 80,000 members.

If you think that you could handle the requirements of an engineering program or an E.T. program equally well, and you are willing to spend four years in school, it is probably best to pursue the engineering degree. An engineering degree provides more options and can put you on a faster track for management positions. Most employers reviewing comparable students, one with an E.T. degree and one with an engineering degree, will take the engineer first. Pay is usually less, by about $2,000–$3,000 annually for starting E.T. grads with four-year degrees than for B.S. engineers. That difference is sometimes made up over time as the E.T. graduates gain experience with their employers.

TYPES OF E.T. DEGREES

The range of E.T. degrees is, in some ways, even broader than that of engineers. Some of the degrees overlap—for instance, to work in the automotive industry one can obtain a mechanical engineering technology degree and also an automotive technology degree. ABET accredits most E.T. programs, but often they are created more rapidly than the accreditation program can adjust. Therefore, some programs are graduating students before the full accreditation is garnered. The following table is a list of E.T. degrees. If you are limited to a certain area or number of schools, it is wise to check with those schools to see if other E.T. degrees are offered.

Engineering Technology Degree Titles

air-conditioning, heating, and refrigeration
aeronautical, astronautical
architectural
automotive
chemical
civil, highway, surveying
construction
computer
drafting, design
electrical
electromechanical
electronic
environmental
engineering/science
general
industrial and manufacturing
marine
mechanical
mining, metallurgical
materials
nuclear
welding

As you can see, the list is a long one, and new degrees are continually being added. There is no definitive count of how many E.T. graduates there are, but the Engineering Manpower Commission, which provides the most comprehensive accounting, recorded about 12,000 four-year graduates in 1992 and a comparable number of two-year graduates. Thus, for every E.T. graduate, there are about five engineering graduates, including those with master's and Ph.D. degrees.

WORK ENVIRONMENTS

Although engineers work predominantly in manufacturing or construction, they also find opportunities in consulting, business services, and government. Most E.T. graduates work in manufacturing and construction exclusively. With the practical training they receive, E.T. graduates are well equipped to perform many functions.

Production and maintenance. E.T. graduates can move off the campus and onto the factory floor or construction site as supervisors, maintenance specialists, instrumentation technicians, and quality-control managers. American manufacturers are straining mightily to upgrade the quality and cost-effectiveness of their production lines, and more engineering technologists are being hired in this effort.

Technical service. When a manufacturer sells a particularly complex piece of equipment, such as a large computer or the instrumentation system for a factory line, many technicians and technologists go along to help with the installation. After the system is up and running, most sales contracts call for the supplier to provide regular maintenance. Sometimes the equipment is so complex and the maintenance needs so steady, that the supplier stations the maintenance technologist at the installation site permanently. However, technical service can also require lots of travel. For some specialists it offers a way of working part-time at a substantial salary because they are on call and work only when needed.

Quality analysis and laboratory services. Most production lines require constant checking to make sure that products are being put together correctly. In addition, there are a sizable number of independent laboratories that offer analytical services to check on the quality, consistency, and performance of products. These efforts require sophisticated instruments and good judgment, and E.T. graduates are often called on to perform the task.

These are the main areas of employment, but they aren't meant to be exclusive. Inspectors (for private companies or for government agencies), testing and quality-control specialists, and those working in the operation of complex machinery, recordkeeping, surveying or mapping for construction companies, research assistance, and a wide variety of technical support positions are all very real possibilities. Getting other types of work, or going on into design or other areas traditionally thought to be the province of engineers, is up to the initiative of the engineering technologist.

Special mention deserves to be made with regard to computers and computer technology. The range of users of computing equipment is growing so rapidly that it is still possible to get a job as a computer specialist with little formal training in computer science. You can become a technologist by gaining experience on computers either through school or simply by working with computers wherever you can. Of course, the more training you get, the better your career prospects will be.

There is no question that the two-year associate degree is significantly less comprehensive than four-year degrees, whether it is four years of ET or four years of engineering. But the advantage of the two-year degree is that you can get a job that pays fairly well ($20,000 per year or better) with the option of continuing your education to get a full four-year degree. It may also be possible to find an employer who will help fund the continuation of your education.

CHAPTER 7

ENGINEERING EDUCATION

When you make plans to attend college and study engineering, several key factors should influence your choices:

- the type of engineering program you will enter
- the quality of the school
- the prospects for employment or advanced schooling after graduation
- your personal preferences in living and learning

This chapter will help you begin investigating these issues. The next section will outline the types of engineering programs available and discuss the other points mentioned above. Keep in mind that the most important decision is to study engineering. The engineering program, the school you choose, and the other factors are of lesser importance. Once you get started on the road to an engineering career, you can change directions in terms of schooling, employment, or long-term career goals. The most important step is the first one—choosing an engineering education.

THE ENGINEERING CURRICULUM

Nearly all engineering programs begin with introductory courses in physics and chemistry and a math course geared to the student's incoming level of education. Usually the first math courses are two terms of calculus. Some students, already well trained in calculus, take the next level up, which could range from linear algebra to differential equations. Because the use of statistics in quality control is now so critical to manufacturing, many engineering students should take this course as well. Other students, not being well prepared in mathematics, may need a precalculus course, followed by the other math courses.

By spring, the college freshman has made a preliminary choice for an engineering program. In the sophomore year, the student begins taking courses in that engineering program. There is usually an introductory course that deals with the general principles of the engineering discipline. This class may be offered concurrently with other specific engineering courses. Another level of science courses is also taken, which might be more chemistry or physics or courses such as computer science, geology, or biology.

In the junior year, a large part of the curriculum is taken up with engineering courses. By now the math requirements are usually fulfilled. Often these courses have substantial laboratory requirements—three or six hours per week.

In the senior year, the final courses of the undergraduate program are taken. Usually these courses will have a heavy orientation toward design. Students might be assigned courses in which they design actual objects or systems: aircraft wings, chemical manufacturing processes, building construction specifications.

Along the way most colleges specify a minimum set of electives in technical and nontechnical areas. Although engineering students are traditionally thought of as avoiding literature or art courses, these courses are heavily represented. Writing and communication skills are also emphasized, both in nontechnical electives and in engineering courses. For example, in their junior-year laboratory courses, students might be required to present the results of an experimental program orally before the class. Studying a foreign language is usually not a requirement, but many engineering students find it desirable to do so, especially if they plan on advanced schooling or a career that involves foreign travel.

Technical Electives

It is with the technical electives that students have the greatest opportunities to tailor their engineering education. Some schools have formal college major/college minor structures, in which a student majors in one topic while fulfilling a set of requirements for a minor. Some students find it desirable to carry a double major. By choosing elective courses carefully, one can fulfill the course requirements for two separate departments.

More often, though, it is through these technical electives that an engineering student directs his or her training to specific engineering specialties. Civil engineers, for example, can choose to specialize in building construction, environmental work, or materials design through the courses they choose. Mechanical engineers can specialize in aerospace or microelectronics. Electrical engineers can specialize in circuit design, power systems, or industrial process control.

CHOOSING A SCHOOL

There are about a thousand colleges and universities across the country; about three hundred of these have one or several engineering or engineering technology programs. Which school should you choose?

The first step is to make decisions about your personal preferences. You may want to attend a school in the immediate vicinity of your home or a school that is far away. You may want a school in the country or one in an urban setting. Obviously the expense of the school is a major factor, and public universities are usually considerably less expensive than private schools. Once you make some of these decisions, you are then ready to consider the potential advantages of specific schools.

Types of Schools

Although all schools consider themselves unique in some way, there are four main types of engineering schools. These are 1) research universities, 2) engineering schools, 3) state universities, 4) private schools.

Research universities have large, comprehensive science and engineering departments. They also have graduate-level programs that attract funding from the federal government and private industry. The top researchers in the country teach at these schools, although they usually teach only graduate-level students, not undergraduates.

The engineering universities are those whose student body is dominated by the engineering program. Many of these schools include the words "institute of technology" in their name. There are a number of liberal arts departments, but the school's focus is on engineering. There are many advantages to attending a school with a large number of fellow engineers, but that is also its main drawback. Some students simply don't want to spend all their time at college with students in the same basic program.

The third type of school, the state university, is the source of most engineering graduates. Many of these schools are land-grant colleges founded as the American West began to open up in the late nineteenth century. They have tens of thousands of students of all types. The schools are usually affordable, especially for residents of that state, and the quality of education can be high. Some state universities are major research powerhouses in their own right.

The fourth category, private colleges and universities, has the widest range of capabilities and quality. Some are quite good; some have only a few types of engineering programs. Some are tailored to the employment needs of major corporations in the school's area.

Again, a certain amount of personal preference enters into the college-choice question. Would you prefer to attend a large school or a small one? Do you want a school where most of your peers are also engineering students or one where they are not? Would you enjoy a school where laboratories and facilities are not so large but where teachers pay great personal attention to their students?

Other Selection Criteria

All schools that offer accredited engineering degrees must meet the minimum requirements of the Accreditation Board of Engineering and Technology (ABET). Check that accreditation is in order at the school you might attend. Beyond that, check the quality and availability of laboratories and computer equipment. Most schools across the country have had tremendous difficulty keeping up-to-date labs and computer systems because of the high cost of such equipment. But, by developing formal or informal alliances with corporate sponsors, many schools have been able to upgrade their laboratories and computer systems.

The status and reputation of the engineering professors themselves is a difficult issue to address. A very prestigious research university may have top engineers on the faculty, but their interactions with undergraduate students may be minimal or nonexistent. One thing to check closely is how many of the courses are actually taught by a full professor, versus the use of teaching assistants.

The rest of the school, outside of the engineering program, is also important, since many of your courses will be held there. Compare the size and types of liberal arts programs at different schools. Find out how well the engineering students are integrated into the overall student body. At some schools the engineering departments tend to be isolated from the rest of the school (which can have advantages and disadvantages of its own).

A final, but very important, element to check in undergraduate facilities is the quality of the placement office. Some schools have a room with a few books; this is less than minimally acceptable. Many have an office where recruiters from corporations or the public sector visit to interview seniors. These interviews may be supplemented with training and preparation for writing résumés and conducting interviews. Compare the number and quality of firms represented at the placement offices at each school. Check also how successful the school is in placing B.S.E. graduates in postgraduate programs.

ALTERNATIVES TO THE FOUR-YEAR PROGRAM

Most high school graduates going to college plan to attend for four years, take summers off, and graduate. However, there are a variety of alternatives. One of the more important of these for engineering students is the co-op program. Co-op (short for cooperative) is a method of paying for school while you are attending and getting training that might be directly related to your ultimate career. Co-ops are usually arranged so that the student graduates with a B.S.E. degree in five years rather than four. Usually each term is alternated with a term of working with local employers (including summers). The work can range from technician's tasks to being a "junior engineer" who assists working engineers in their tasks.

Co-ops are valuable for engineering students because they can provide training similar to the work that full-fledged engineers perform. Many schools have provisions for co-op education; there are also several where the co-op program is stressed.

Another type of program that some students use is a "3-2" program leading to a master's degree. This often exists at schools that, individually, do not have a fully accredited engineering curriculum. The school accepts students who take all the preparatory courses for an engineering degree during the first three years (i.e., all the math, science, and liberal arts electives). The student then attends another school where the necessary engineering courses are taken.

The P.E. License

A professional engineer's (P.E.) license has the same function as an M.D. title for medical doctors or LL.D or J.D. for lawyers. To be a professional engineer, one must pass an examination on engineering fundamentals, usually taken at the end of one's undergraduate program or immediately after. After passing this exam the student becomes an "engineer in training" (E.I.T.). Then, after being employed as an engineer for several years (usually a minimum of three), the E.I.T. takes a second exam that tests general knowledge of engineering practices. Some schools make provisions in the senior-year curriculum for the E.I.T. exam.

Professional licensing has some practical value and much symbolic value. For certain types of engineers, such as construction specialists who handle public-works projects, a license is necessary. Someone at the firm doing the work must be licensed. His or her professional seal on the final blueprints signifies that a fully qualified professional engineer has examined the plans and approved them. Obviously if you are going to work on your own in public-works construction, you need your P.E. license. If you join a larger firm, with P.E.s already on staff, you may not need the license, but the firm may want you to get one as soon as you can.

The P.E. license also has symbolic value. Engineers who have one show that they are serious about their work and consider themselves true professionals. There are a few situations when having the P.E. license results in slightly higher pay, but this is the exception rather than the rule.

Some people view engineers with a P.E. license as the only true engineers. The importance of the license is a matter of opinion. However, most corporations do not require their engineers to be licensed. Thus only about 15 percent of working engineers have one. P.E. licensing is administered by state organizations that are banded together as the National Council of Engineering Examiners. There is also another organization, the National Society of Professional Engineers, all of whose members are licensed.

FINANCIAL AID

Financing should not be an impossible obstacle to obtaining an engineering degree. Although the costs of a college education are high and growing higher every year, scholarships, grants, work-study programs, and other forms of financial aid are available. The better your academic record, the more opportunities appear. Women and minority students have many sources of funding for engineering education in addition to the ones available to all students.

Engineering usually has a special status in the view of the federal government because of its importance to the manufacturing capability of the country and the need for supplying high-tech military equipment. At this time there are no special financial programs from the federal government for engineering students at the undergraduate level. At the master's and higher levels, however, there is funding from the National Science Foundation and other sources to help engineers continue their education.

Private industry helps out by sponsoring a variety of scholarships for engineers. A special variation of the work-study program is the availability of summer work for college juniors and, sometimes, sophomores. Many companies use these programs as a means of introducing the company to potential recruits. The availability of these temporary jobs depends on the state of the economy and the demand for engineering graduates. When the number of engineering students falls, companies try to increase the summer openings in order to attract more students to the various engineering disciplines.

Check in reference books, with high school advisers, and with the financial-aid counselors at the college you would like to attend for information on scholarships, work-study, and summer jobs. Don't give up until you have obtained the financing you need to attend school.

WHAT TO DO NOW

Assuming that you are a sophomore or junior in high school, there are several things you can do now to prepare for studying engineering.

An obvious first step is to take as many science and math courses as you can. You can't overprepare in this sense because all the math and science you learn will be applied in an engineering program. You should do well in these courses, although there may be many valid reasons for getting less than As all along the way. If you don't do well, ask yourself if you gave the courses your best effort, or if there were other reasons you did not excel. Be confident of your abilities in math and science.

There is an educational association sponsored by the professional engineering societies that offers a big boost to high school students considering studying engineering. This association is JETS—the Junior Engineering Technical Society. JETS sponsors academic competitions, design contests, workshops, and other activities that help high school students learn about engineering. One such program, the Tests of Engineering Aptitude, Mathematics, and Science (TEAMS) competition, encourages higher order thinking, leadership skills, academic excellence, and working successfully as a team to solve problems faced by engineers and scientists. JETS also sponsors the National Engineering Aptitude Search, a guidance test that helps students gauge how well-prepared they are for studying engineering. Its guidance program offers activities and materials that can be used in math, science, and technology clubs at high schools.

JETS also helps organize contact between working engineers and high school students contemplating an engineering career. Contacting a working engineer can be a valuable help in determining what kind of engineering work you might want to do. Ask family and friends for a reference to engineers. Then call or write to them, asking for an interview and perhaps a tour of their company's facilities. Or, call or write directly to manufacturing or construction companies in your area, seeking the same things. Getting this exposure will help you find out where your interests are and will provide a role model for carrying you through your engineering program. Good luck!

APPENDIX A

RECOMMENDED READING

Basta, Nicholas. *Environmental Jobs for Engineers and Scientists.* New York: Wiley, 1992.

CEIP Fund. *Complete Guide to Environmental Careers.* Washington, DC: Island Press, 1993.

Florman, Samuel C. *Blaming Technology.* New York: St. Martin's Press, 1981.

————. *The Civilized Engineer.* New York: St. Martin's Press, 1987.

————. *The Existential Pleasures of Engineering.* New York: St. Martin's Press, 1976.

McMahon, A. Michal. *The Making of a Profession: A Century of Electrical Engineering in America.* New York: IEEE Press, 1984.

Manes, Stephen and Paul Andrews. *Gates: How Microsoft's Mogul Reinvented an Industry.* New York: Doubleday, 1993.

Reynolds, Terry S. *Seventy-Five Years of Progress: A History of the American Institute of Chemical Engineers 1908-1983.* New York: American Institute of Chemical Engineers, 1983.

Sinclair, Bruce. *A Centennial History of the American Society of Mechanical Engineers 1880-1980.* Toronto: Toronto University Press, 1980.

Stewart, Robert E. *Seven Decades that Changed America: A History of the American Society of Agricultural Engineers.* St. Joseph, Mich.: ASAE, 1979.

ENGINEERING AND TECHNOLOGY ASSOCIATIONS

Following is a list of associations mentioned in this book, along with a number of related groups. This list by no means represents the total number of engineering-related associations, which runs into the hundreds. A good reference for further inquiry is the *1989 Directory of Engineering Societies,* edited by Gordon Davis and published by the American Association of Engineering Societies. See Appendix A for details.

Most of these associations provide educational or career-planning literature. The Junior Engineering Technical Society, listed below, has brief pamphlets describing most engineering fields.

Alliance for Engineering in Medicine and Biology
1101 Connecticut Avenue, NW
Washington, DC 20036

American Academy of Environmental Engineers
132 Holiday Court, Suite 206
Annapolis, MD 21401

American Institute of Aeronautics and Astronautics
370 L'Enfant Promenade, SW
Washington, DC 20024

American Institute of Chemical Engineers
345 East 47th Street
New York, NY 10017

American Institute of Mining, Metallurgical, and
    Petroleum Engineers (AIME)
    345 East 47th Street
    New York, NY 10017

American Institute of Physics
    335 East 45th Street
    New York, NY 10017

American Institute of Plant Engineers
    3975 Erie Avenue
    Cincinnati, OH 45208

American Nuclear Society
    555 North Kensington Avenue
    La Grange Park, IL 60525

American Society of Agricultural Engineers
    2950 Niles Road
    St. Joseph, MI 49085

American Society of Civil Engineers
    Student Services Department
    345 East 47th Street
    New York, NY 10017

American Society of Heating, Refrigerating, and Air-Conditioning
    Engineers, Inc. (ASHRAE)
    1791 Tullie Circle, NE
    Atlanta, GA 30329

American Society of Mechanical Engineers
    345 East 47th Street
    New York, NY 10017

American Society of Naval Engineers
    1452 Duke Street
    Alexandria, VA 22314

American Society of Safety Engineers
    1800 E. Oakton Street
    Des Plaines, IL 60018

ASM International (American Society for Metals)
    9639 Kinsmen Road
    Metals Park, OH 44073

Institute of Electrical and Electronics Engineers, Inc.
    345 East 47th Street
    New York, NY 10017

Institute of Industrial Engineers
    25 Technology Park
    Norcross, GA 30092

Instrument Society of America
    Education Services
    67 Alexander Drive
    P.O. Box 12277
    Research Triangle Park, NC 27709

Junior Engineering Technical Society (JETS)
    1420 King Street, Suite 405
    Alexandria, VA 22314-2715

The Metallurgical Society
    420 Commonwealth Drive
    Warrendale, PA 15086

National Action Council for Minorities in Engineering
    3 West 35th Street
    New York, NY 10001

National Institute of Ceramic Engineers
    65 Ceramic Drive
    Columbus, OH 43214

National Society of Professional Engineers (NSPE)
    1420 King Street
    Alexandria, VA 22314

Operations Research Society of America
    Mt. Royal & Guilford Avenue
    Baltimore, MD 21202

Optical Society of America
    1816 Jefferson Place, NW
    Washington, DC 20036

Society of Automotive Engineers
    400 Commonwealth Drive
    Warrendale, PA 15096

Society of Fire Protection Engineers
    60 Batterymarch Street
    Boston, MA 02110

Society of Manufacturing Engineers
    One SME Drive
    Dearborn, MI 48121

Society of Naval Architects and Marine Engineers
    One World Trade Center, Suite 1369
    New York, NY 10048

Society of Plastics Engineers
    14 Fairchild Drive
    Brookfield, CT 06804

Society of Women Engineers
    345 East 47th Street
    New York, NY 10017

A complete list of titles in our extensive *Opportunities* series

OPPORTUNITIES IN

Accounting
Acting
Advertising
Aerospace
Airline
Animal & Pet Care
Architecture
Automotive Service
Banking
Beauty Culture
Biological Sciences
Biotechnology
Broadcasting
Building Construction Trades
Business Communication
Business Management
Cable Television
CAD/CAM
Carpentry
Chemistry
Child Care
Chiropractic
Civil Engineering
Cleaning Service
Commercial Art & Graphic Design
Computer Maintenance
Computer Science
Counseling & Development
Crafts
Culinary
Customer Service
Data Processing
Dental Care
Desktop Publishing
Direct Marketing
Drafting
Electrical Trades
Electronics
Energy
Engineering
Engineering Technology
Environmental
Eye Care
Farming and Agriculture
Fashion
Fast Food
Federal Government
Film
Financial

Fire Protection Services
Fitness
Food Services
Foreign Language
Forestry
Franchising
Gerontology & Aging Services
Health & Medical
Heating, Ventilation, Air Conditioning, and Refrigeration
High Tech
Home Economics
Homecare Services
Horticulture
Hospital Administration
Hotel & Motel Management
Human Resource Management
Information Systems
Installation & Repair
Insurance
Interior Design & Decorating
International Business
Journalism
Laser Technology
Law
Law Enforcement & Criminal Justice
Library & Information Science
Machine Trades
Marine & Maritime
Marketing
Masonry
Medical Imaging
Medical Technology
Mental Health
Metalworking
Military
Modeling
Music
Nonprofit Organizations
Nursing
Nutrition
Occupational Therapy
Office Occupations
Paralegal
Paramedical
Part-time & Summer Jobs
Performing Arts
Petroleum
Pharmacy
Photography

Physical Therapy
Physician
Physician Assistant
Plastics
Plumbing & Pipe Fitting
Postal Service
Printing
Property Management
Psychology
Public Health
Public Relations
Publishing
Purchasing
Real Estate
Recreation & Leisure
Religious Service
Restaurant
Retailing
Robotics
Sales
Secretarial
Social Science
Social Work
Special Education
Speech-Language Pathology
Sports & Athletics
Sports Medicine
State & Local Government
Teaching
Teaching English to Speakers of Other Languages
Technical Writing & Communications
Telecommunications
Telemarketing
Television & Video
Theatrical Design & Production
Tool & Die
Transportation
Travel
Trucking
Veterinary Medicine
Visual Arts
Vocational & Technical
Warehousing
Waste Management
Welding
Word Processing
Writing
Your Own Service Business

VGM Career Horizons
a division of *NTC Publishing Group*
4255 West Touhy Avenue
Lincolnwood, Illinois 60646–1975